Semiconductors
from Book to Breadboard

Complete Textbook/Lab Manual

Semiconductors from Book to Breadboard

Complete Textbook/Lab Manual

Kevin A. McGowan

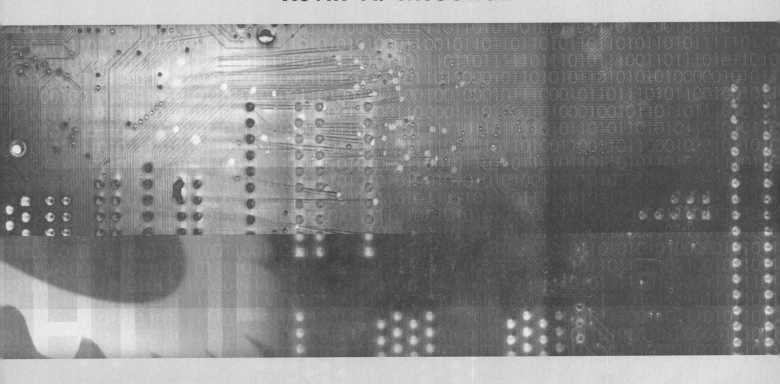

DELMAR
CENGAGE Learning

Australia • Brazil • Japan • Korea • Mexico • Singapore • Spain • United Kingdom • United States

DELMAR
CENGAGE Learning™

Semiconductors from Book to Breadboard: Complete Textbook/ Lab Manual

Kevin A. McGowan

Vice President, Editorial: Dave Garza

Director of Learning Solutions: Sandy Clark

Acquisitions Editor: Stacy Masucci

Managing Editor: Larry Main

Senior Product Manager: John Fisher

Editorial Assistant: Andrea Timpano

Vice President, Marketing: Jennifer Baker

Marketing Director: Deborah Yarnell

Senior Marketing Manager: Erin Brennan

Marketing Coordinator: Jillian Borden

Senior Production Director: Wendy Troeger

Production Manager: Mark Bernard

Content Project Manager: Barbara LeFleur

Production Technology Assistant: Emily Gross

Senior Art Director: David Arsenault

Technology Project Manager: Joe Pliss

Cover Image: © Tim Conners

Interior Image Credits: All photos by Tracy Grace Leleux unless otherwise noted

For product information and technology assistance, contact us at **Cengage Learning Customer & Sales Support, 1-800-354-9706**

For permission to use material from this text or product, submit all requests online at **www.cengage.com/permissions**. Further permissions questions can be e-mailed to **permissionrequest@cengage.com**

Library of Congress Control Number: 2011926108

ISBN-13: 978-1-111-31387-6

ISBN-10: 1-111-31387-3

Delmar
5 Maxwell Drive
Clifton Park, NY 12065-2919
USA

Cengage Learning is a leading provider of customized learning solutions with office locations around the globe, including Singapore, the United Kingdom, Australia, Mexico, Brazil, and Japan. Locate your local office at: **international.cengage.com/region**

Cengage Learning products are represented in Canada by Nelson Education, Ltd.

To learn more about Delmar, visit **www.cengage.com/delmar**

Purchase any of our products at your local college store or at our preferred online store **www.cengagebrain.com**

Notice to the Reader

Publisher does not warrant or guarantee any of the products described herein or perform any independent analysis in connection with any of the product information contained herein. Publisher does not assume, and expressly disclaims, any obligation to obtain and include information other than that provided to it by the manufacturer. The reader is expressly warned to consider and adopt all safety precautions that might be indicated by the activities described herein and to avoid all potential hazards. By following the instructions contained herein, the reader willingly assumes all risks in connection with such instructions. The publisher makes no representations or warranties of any kind, including but not limited to, the warranties of fitness for particular purpose or merchantability, nor are any such representations implied with respect to the material set forth herein, and the publisher takes no responsibility with respect to such material. The publisher shall not be liable for any special, consequential, or exemplary damages resulting, in whole or part, from the readers' use of, or reliance upon, this material.

Printed in the United States of America
2 3 4 5 6 7 8   20 19 18 17 16

Table of Contents

SECTION 2
Transistors

CHAPTER 4
Bipolar Junction Transistors (BJTs)

CHAPTER 5
Amplifier Configurations

CHAPTER 6
Amplifier Classes

CHAPTER 7
Field Effect Transistors (FETs)

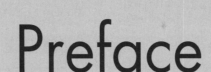

Preface

The idea for this book stemmed from the need to teach the concepts and skills of semiconductors/solid-state devices in less than 90 classroom hours. My 12 years of teaching electronics courses have been a process of continual refinement, and being a working technician and teacher, I know what works for students and what skills are needed in the workplace.

Student evaluations commonly stress the need for more hands-on training; however, hands-on experiences should not exist alone, just as theory without practice has its limits. Electronic technicians work with their heads *and* their hands, and it is a fact of education that hands-on activities dramatically improve the learning process. My colleagues and students agree that semiconductor theory is grounded in physics and chemistry and that it involves abstract concepts. The most successful approach to teaching Semiconductors—and the purpose of this book—is to explain these concepts in simple terms and then provide *multiple reinforcements.*

I wrote this book with these two processes in mind: simplifying concepts and providing multiple reinforcements. Throughout the chapters in this book, self-contained blocks of instruction use a lesson-to-lab format so that within one classroom period (one to two hours), a student can progress easily from concept to hands-on activities involving semiconductors. Because students need to touch and test a semiconductor device, such as a diode, on Day One, lab assignments are included in nearly every section. *Thus, the textbook is a combination of concepts and labs.* Concepts are explained in clear and simple terms (without dumbing down or diluting the curriculum), and common electronic components are used for labs, thus minimizing inventory. Also, each lab activity has already been tested multiple times in the classroom, thus eliminating procedural errors or false results.

The book is designed for students in 12 month electrical/electronics vocational diploma programs and two-year or four-year electronics technology degree programs. It will satisfy the requirements for course names such as Semiconductors, Solid-State Devices, Semiconductor Fundamentals, and Transistor Circuits. In most cases, the objectives for all lessons can be satisfied in a course that covers a time frame of 45–90 hours. A student should have a background in Direct Current (DC) and Alternating Current (AC).

I use a five-part approach for teaching Semiconductors that involves the introduction and discussion of a concept, two to three reinforcements, and then an evaluation:

1) PowerPoints (included on the Instructor Companion Website) and a dry erase board to teach a concept.

2) Solid State Challenges™ (available from www.etcai.com if your classroom does not already have them).

3) Circuit simulation software such as Multisim®, Electronics Workbench®, or SPICE that is used to design and test circuits. (If your classroom computers do not have any the above software, many on-line sites offer a free download.)

4) Hands-on lab assignments (included at the end of most lessons in the textbook).

5) Test generator with an extensive test bank for each chapter (included on the Instructor Companion Website).

Of course, this is a suggested approach. Each class and lesson is unique and requires its own type of reinforcement. An instructor may use any combination of the five techniques and resources to reach his or her objectives.

The Solid State Challenges™ are comprised of 12 interactive software lessons. To enhance learning, the instructor should select, view, and practice a desired Challenge and then include it as part of an applicable lesson. The Solid State Challenges™ include the following:

1) Basic Diode Testing.

2) Basic Transistor Testing.

3) Common-Emitter Amplifiers One (Base Bias).

4) Common-Emitter Amplifiers Two (Voltage Divider Bias).

5) Common-Emitter Amplifiers Three (with Emitter Bypass).

6) Common-Collector Amplifiers.

7) Common-Base Amplifiers.

8) Class B Power Amplifiers One.

9) Class B Power Amplifiers Two.

10) Class B Power Amplifiers Three (with Darlington).

11) Troubleshooting Simple Amplifiers.

12) Troubleshooting Common-Emitter Amplifiers.

As mentioned above, the Solid State Challenges™ are available from www.etcai.com.

Supplements

INSTRUCTOR COMPANION SITE

The Instructor Companion Site is an educational resource that creates a truly electronic classroom. It contains tools and instructional resources that will enrich your classroom and make your preparation time shorter. The elements of the *Instructor Companion Site* link directly to the text and tie together to provide a unified instructional system.

Features contained in the *Instructor Companion Site* include:

• Instructor's Guide: Contains answers to all end-of-chapter review questions and lab solutions.

• Chapter PowerPoint Presentations: Provide the basis for lecture outlines that help you present concepts and material. Key points and concepts can be graphically highlighted for student retention.

• Image Gallery: Images from the textbook allow you to customize PowerPoint presentations or to use them as transparency masters. This is a quick and easy tool for enhancing teaching and research projects.

Instructor Site

Contact Delmar Cengage Learning or your local sales representative to obtain an instructor account.

ACCESSING AN INSTRUCTOR COMPANION WEBSITE FROM SSO FRONT DOOR

1) GO TO: http://login.cengage.com and login using the Instructor email address and password.

2) ENTER author, title, or ISBN in the **Add a title to your bookshelf** search box, and CLICK on **Search** button.

3) CLICK **Add to My Bookshelf** to add Instructor Resources.

4) At the Product page click on the **Instructor Companion site** link.

NEW USERS

If you're new to Cengage.com and do not have a password, contact your sales representative.

ABOUT THE AUTHOR

Formerly a professional technician for a paper-processing plant and an oilfield service company, Kevin A. McGowan has been teaching electronics and computer-related courses for over a decade. He is a Journeyman Certified Electronics Technician with the International Society of Certified Electronics Technicians (I.S.C.E.T.), is A+ Certified by CompTIA, and holds an F.C.C. General Radiotelephone Operator License. Mr. McGowan earned an A.S. in Electrical Engineering Technology at Penn State University, a B.S. in English at Austin Peay State University, an M.A. in English at McNeese State University, and an M.Ed. in Administration and Supervision at the University of Louisiana, Lafayette.

ACKNOWLEDGMENTS

I would like to thank the following people for helping make this book possible: James Corrick, copyediting; James Cox, for his insight regarding voltage multipliers, voltage dividers, and oscillators; Dawn Daugherty, who guided me through the proposal process; John Fisher, for managing the integration of words, photos, and art; Chris Kilian, control technology extraordinaire; Maria Klimek, marketing guru; Kari LaFontaine, for rapid mailing of books and things and for introducing me to the Cengage family; Nithya Kuppuraj, who converted the manuscript to a book; Barbara LeFleur, coordinator of content; Tracy Grace Leleux, for her photography expertise; Electronics teachers at Remington College, Lafayette, Louisiana, past and present; Stacy Masucci, who gave the green light on the whole project; Maria Ponce de Leon, for awesome editing of the entire manuscript; Andrea Timpano, who handled forms and artifacts; Mike Brumbach, Mark C. Henton, Sr., Marvin E. Moak, and Ken Warfield, for technical editing and suggestions; and all the electronic giants, past and present.

MATERIALS, EQUIPMENT, AND PARTS LIST FOR ALL LAB ACTIVITIES

The materials, equipment, and parts list includes components needed for all lab activities throughout the book. A separate materials, equipment, and parts list is also provided at the beginning of each chapter for the lab activities within that chapter. In many cases, *equivalent parts* will suffice, providing they meet similar voltage, current, and power ratings. Note: Quantity is one per component unless otherwise noted. Special operating characteristics are also listed when applicable.

Manuals and Software:

- *NTE Semiconductor Technical Guide and Cross Reference* catalog, NTE QUICKCross™ software, or Internet access, www.nteinc.com
- PC w/Multisim® 7, Electronics Workbench®, or SPICE.
- PC w/Solid State Challenges™.

Test and Measuring Instruments:

- Digital Multimeter (DMM) with test leads.
- Dual trace oscilloscope with/BNC-to-alligator leads.
- 5 V DC voltage source.
- 12 V DC voltage source.
- Variable positive DC voltage source, 0 V to 20 V.
- Variable negative DC voltage source, 0 V to 20 V.
- 6.3 V AC voltage source.
- 12.6 V AC voltage source w/center-tap.
- Function generator, 1 Hz to 30 kHz.

Miscellaneous:

- Breadboard and connecting wires.
- Grounded wrist strap or antistatic mat.
- SPST switch (3).

Integrated Circuits (ICs):

- LM741CN, UA741CN, or equivalent op amp.
- 4N25 or 4N35 optoisolator.
- Optional: IC bridge rectifier such as DB102 (NTE 5332), DF02 (NTE 5332), *or* one that can handle 1 A for safety sake.

Transistors:

- BJTs: 2N2219 (NPN), 2N3053 (NPN), 2N3055 (NPN), 2N3904 (NPN), 2N3906 (PNP), and 2N6373 (NPN).
- JFETs (all N-channel): 2N5457, 2N5459, and 2N5485.
- D-MOSFETs: DN1509.
- E-MOSFETs (all N-channel): BS170, IRF510, IRF620, and 2N7000.

Diodes:

- General-purpose rectifiers, silicon: 1N4001 (4), 1N4002, 1N4004, and 1N4007.
- General-purpose rectifiers, germanium: 1N34, 1N60, and 1N277.
- Fast-switching diodes: 1N914 and 1N4148 (2).
- Zener diodes: 1N4728 (3.3 V_Z), 1N4733A (5.1 V_Z), 1N4742A (12 V_Z), 1N4743 (13 V_Z), 1N5235B (6.8 V_Z), 1N5240B (10 V_Z), and 1N5248 (18 V_Z).
- Silicon-controlled rectifiers: C106-series such as C106B or C106D; T106-series such as T106B1.
- Two-terminal LEDs.
- Common-anode, seven-segment, single-digit LED display, MAN4610A or equivalent.

Capacitors (50 V DC rating recommended):

- 680 pF, 0.01 µF, 0.1 µF (3), 0.22 µF, 1 µF (3), 10 µF (2), 22 µF, 47 µF (2), 100 µF, and 220 µF.

Inductors (Coils):

- 4.7 mH (2), 10 mH, 47 mH, and 100 mH.

Fixed resistors (¼ watt or greater recommended):

- 33 Ω, 100 Ω, 150 Ω, 180 Ω, 220 Ω, 270 Ω (4), 330 Ω, 470 Ω, 680 Ω, 820 Ω, 1 kΩ (2), 1.2 kΩ, 1.5 kΩ, 1.8 kΩ (2), 2.2 kΩ (2), 2.7 kΩ, 3.9 kΩ, 4 kΩ, 4.3 kΩ, 5 kΩ, 6.8 kΩ, 8.2 kΩ, 10 kΩ (2), 12 kΩ, 15 kΩ, 20 kΩ, 22 kΩ, 33 kΩ, 39 kΩ, 68 kΩ, 82 kΩ, 100 kΩ, 220 kΩ, 820 kΩ, 1 MΩ, and 1.2 MΩ.

Diodes

CHAPTER **1**

Semiconductor Principles and the PN Junction Diode

OBJECTIVES *Upon completion of this chapter, you should be able to:*

- Distinguish between solid-state technology and vacuum tube technology.
- List the advantages and disadvantages of semiconductors.
- Explain how a depletion region is created in the manufacture of a diode and why the depletion region is important to electronic technicians.
- Compare the barrier voltage of a germanium diode to that of a silicon diode.

- Draw and label the schematic symbol for a PN junction diode.
- Measure a diode with a digital multimeter (DMM) and determine if the diode is good or bad.
- Use a semiconductor guide to identify replacement diodes.
- Build a diode circuit to calculate and measure voltages and total circuit current.

MATERIALS, EQUIPMENT, AND PARTS

Materials, equipment, and parts needed for the lab experiments in this chapter are listed below:

- *NTE Semiconductor Technical Guide and Cross Reference* catalog, NTE QUICKCross™ software, or Internet access, www.nteinc.com
- PC w/Multisim® 7, Electronics Workbench®, or SPICE.
- Digital Multimeter (DMM) with test leads.
- 5 V DC power source.

- Breadboard and connecting wires.
- Assortment of silicon diodes such as 1N4001, 1N4002, 1N4004, and 1N4007.
- Assortment of germanium diodes such as 1N34, 1N60, and 1N277.

- Assortment of fast-switching diodes such as 1N914 and 1N4148.
- Assortment of Zener diodes such as 1N4728, 1N4743, and 1N5248.
- 1 kΩ fixed resistor.

GLOSSARY OF TERMS

Vacuum tube technology The science concerning a vacuum tube, an electronic device in which electrons (electricity) move through a low-pressure space, or vacuum

Solid-state technology The science involving the movement of electrons (electricity) through a solid piece of semiconductor material, typically silicon or germanium

Depletion region The area between the P-type material and

N-type material of a semiconductor device that lacks electrons

Barrier voltage (V_B) A force between the P-type material and N-type material of a semiconductor device

Forward bias condition A situation that occurs in a semiconductor material when the barrier or depletion region breaks down and current flows through the device

Bias A DC voltage applied to a semiconductor device to turn it on or off

Reverse bias condition A situation that occurs in a semiconductor material when the depletion region increases and no current flows through the device

Rectifier A semiconductor device or circuit that changes AC to pulsating DC

1-1 IT'S A SOLID-STATE WORLD

We live in a solid-state or semiconductor world. Everything from personal computers to calculators to iPods to digital billboards uses solid-state circuitry or semiconductors. In fact, one would be hard-pressed to find some electronic device that doesn't have semiconductor components.

However, it wasn't always that way. From the 1920s to the mid-1950s, radios, TVs, computers, and industrial electronics used **vacuum tube technology**. Figure 1-1 shows a vacuum tube, a 1N4001 PN junction diode, and a 2N3904 transistor.

The vacuum tube looks like an old-fashioned Christmas light bulb. In a vacuum tube, electrons (electricity) move through a low-pressure *space*, or vacuum. Vacuum tubes wasted a lot of power (think of the heat generated by just a 100 watt light bulb), shattered easily since they were encased in glass, frequently had to be replaced, and took up a lot of space. In fact, your grandparents or great-grandparents might have a vacuum tube radio or TV tucked away in the attic or garage. Vacuum tube radios and TVs were so large that they used to be considered furniture. ENIAC, the world's first electronic computer, had nearly 18,000 vacuum tubes and took up roughly 1500 square feet, the size of a modern three-bedroom house.

INTERNET ALERT

Check the website http://inventors.about.com/od/estartinventions/a/Eniac.htm to discover more about vacuum tubes and ENIAC.

Consumer solid-state technology first appeared about 60 years ago with the invention of the transistor and the PN junction diode, which led to the pocket-size transistor radio. In **solid-state technology**, electrons (electricity) move through a *solid* piece of semiconductor material, typically silicon or germanium. Looking again at Figure 1-1, it's obvious that semiconductors such as the PN junction diode and 2N3904 transistor are smaller than their vacuum tube equivalents. Solid-state components have several advantages over vacuum tubes:

- They use less power.
- They are more rugged.
- They last longer.
- They take up less space.

Solid-state devices such as the PN junction diode and the transistor haven't replaced vacuum tubes in every application, however. Vacuum tubes are still used today in very high-power radio and TV transmitters that use thousands of watts in their amplifier output stages. This is because solid-state amplifiers haven't been built yet that can handle such power. Also, many musicians still use vacuum tubes in audio amplifiers to re-create songs with a 1950s and 1960s feel and sound.

The terms "solid-state" and "semiconductor" are often used interchangeably. However, solid-state really refers to the method of transmitting

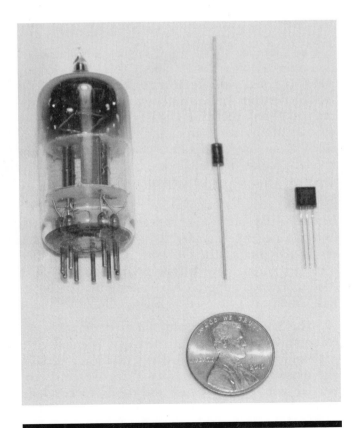

FIGURE 1-1 Vacuum tube, 1N4001 PN junction diode, and 2N3904 transistor (left to right) (Courtesy of Tracy Grace Leleux)

energy, while semiconductor is the material or device the energy travels within. For the remainder of this book, the term semiconductor will be used to refer to the various devices that use solid-state technology, not vacuum tubes.

1-2 THE PN JUNCTION DIODE

The PN junction diode is the simplest and most commonly used semiconductor device. There are hundreds of diodes and countless applications, but a technician will probably encounter only a few basic types, which will be discussed in detail in this chapter and the next. When technicians say the word "diode" by itself, they are referring to the PN junction diode.

The PN junction diode is composed of P-type material such as boron and N-type material such as arsenic that are chemically combined in a silicon (Si) or germanium (Ge) base. The boron atoms of P-type material are lacking electrons, so they have excess "holes" and an overall positive charge. The arsenic atoms of N-type material have excess electrons and an overall negative charge.

Figure 1-2(a) shows a simplified drawing of P-type material and N-type material before they are chemically combined. The P-type material is on the left, and the N-type material is on the right. The vertical blue dotted line separating the P-type and N-type materials is called a junction. (The blue dotted line is used for learning purposes only: you can't split a diode down the middle into P-type and N-type materials.) Each P-type atom is shown as an empty circle, and each N-type atom is represented by a negative sign.

Figure 1-2(b) shows what happens during the manufacturing process. Some electrons from the N-type material on the right cross over to the P-type material on the left and fill the holes of the P-type atoms. Once the electrons fill all the holes along the border of the P-type material, a wall of electrons in the P-type material forms a barrier and prevents any more electrons of the N-type material from crossing over to the P-type material because electrons have a negative charge and repel each other.

Figure 1-2(c) shows the result of the chemical combination of atoms of P-type material and N-type material. There is a "no electron zone" between the P-type material on the left and the

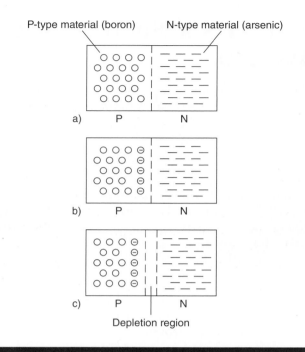

FIGURE 1-2 Creation of depletion region: a) Before electrons and holes combine, b) Electrons filling holes of P-type material, c) After electron-hole combination (© Cengage Learning 2012)

N-type material on the right; this zone is shown by two vertical blue dotted lines. This "no electron zone" is called a **depletion region** because of the lack of electrons in the area between the P-type material and N-type material. The P-type material still has an overall positive charge, and the N-type material still has an overall negative charge. However, the overall charge of the diode itself is *neutral*—you can't get shocked by picking up a diode.

This area or depletion region or junction between the positively charged P-type material and the negatively charged N-type material in a diode *acts* like a little DC battery, and it is called a barrier potential, or **barrier voltage (V$_B$)**. The diode, however, doesn't have a true battery inside it; a diode can't be used as a DC voltage source.

This barrier voltage within a diode opposes any external voltage source. If the diode is made of a germanium base, an applied external force or voltage must be greater than 0.3 V to overcome the barrier voltage and cause the diode to conduct. If the diode is made of a silicon base, an applied external force or voltage must be greater than 0.6 V to overcome the barrier voltage and cause the diode to conduct. The applied external voltage that causes a diode to conduct is called forward bias, and it will be discussed in detail later. For now, *this is what is most important to a technician: a good PN junction diode will measure a voltage drop of @ 0.3 V if it's germanium and @ 0.6 V if it's silicon.* As aspiring technicians, we must be able to measure a diode to tell if it's good or bad. In fact, we have an upcoming lab where we will measure diodes.

Figure 1-3(a) shows a PN junction diode with the part number 1N4007. You can't see the entire number because it wraps around the diode. Most diodes have the prefix 1N. Notice the white stripe on the one end of the diode. Manufacturers often mark the cathode end with a stripe for identification purposes.

Figure 1-3(b) shows the schematic symbol for a PN junction diode. The end of the diode connected to the P-type material is called the anode, which may be labeled with the letter *A* in drawings. The end of the diode connected to the N-type material is called the cathode, which may be labeled with the letter *K* in drawings. The cathode of a diode is usually connected to the negative side or ground in a circuit.

It's important to understand that the schematic symbol does not indicate if a diode is made of germanium (Ge) or silicon (Si). Silicon diodes have almost totally replaced germanium diodes in electronic applications. Although germanium was the first element used in the manufacture of diodes, silicon diodes are easier and cheaper to manufacture, and they can handle more heat. Germanium diodes are still used in some applications, however, where only a small voltage of 0.3 V is needed to turn on the diode, as in a crystal radio set.

INTERNET ALERT

To learn more details about diodes, see the website http://electronics.howstuffworks.com/diode.htm

1-3 Using a DMM to Test a Diode

In Section 1-2, we learned that during the manufacturing process of a diode, a depletion region is created in the area between the P-type material and N-type material. The P-type or anode end of the diode is the positive end, and the N-type or cathode end of the diode is the negative end. We also learned that a force, or barrier voltage, exists between the P-type material and the N-type material: @ 0.3 V for a germanium diode and @ 0.6 V for a silicon diode.

The main tool of the electronic technician is a digital multimeter (DMM). Figure 1-4 shows an RSR™ Model 717 DMM. Most DMMs have a diode setting used to measure both diodes and transistors. In Figure 1-4, you can clearly see the

a)
Courtesy of Tracy Grace Leleux

Anode   Cathode
(A)   (K)

b)   +   ▶| −
© Cengage Learning 2012

FIGURE 1-3 PN junction diode: a) 1N4007 diode and b) Schematic symbol

FIGURE 1-4 RSR™ Model 717 Digital Multimeter on diode setting (Courtesy of Electronix Express/RSR Electronics, Inc.)

selector switch is on the diode setting, which looks like the diode schematic symbol. When you check a diode on the diode setting, the numerical result is in volts, though the meter may *not* show "volts" or the letter *V*.

To check a diode, set your meter's rotary selector switch to the diode setting. Then touch or clip the red (positive) lead of the meter to the anode end of the diode and touch or clip the black (negative) lead of the meter to the cathode (striped) end of the diode. If the diode is germanium, the meter should display from 0.2 to 0.3 (volt) for a good diode; if the diode is silicon, the meter should display between 0.6 and 0.7 (volt) for a good diode. (Actually, some may read *slightly* less or more.) This is called the **forward bias condition**—positive lead of meter to positive (anode) end of diode and negative lead of meter to negative (cathode) end. Forward bias (positive to positive and negative to negative) breaks down the barrier or depletion region in a diode, and current flows through the diode. **Bias** is just another name for a DC voltage.

Diode testing requires a second check. Touch or clip the red (positive) lead of the meter to the cathode (striped) end of the diode, and touch or clip the black (negative) lead of the meter to the anode end of the diode. If the diode is germanium **or** silicon, the meter should display infinite, which may be represented by the infinity symbol (∞) or "OL" (overload) for a good diode. This is called the **reverse bias condition**—positive lead of meter to negative (cathode) end of diode and negative lead of meter to positive (anode) end. Reverse bias (positive to negative and negative to positive) enlarges the depletion region (the "no electron zone") in a diode, and no current flows through the diode.

Thus, a good diode conducts one-half the time—when it's in the forward bias condition. Consequently, a germanium diode must read about 0.2 to 0.3 (volt) in the forward bias condition *and* infinite in the reverse bias condition to be "good." If a germanium diode reads 0.2 to 0.3 in *both* the forward and reverse bias conditions, it is bad (shorted) and must be replaced. Also, if a germanium diode reads infinite in *both* the forward and reverse bias conditions, it is bad (open) and must be replaced. The same holds true for a silicon diode. A silicon diode must read about 0.6 to 0.7 (volt) in the forward bias condition *and* infinite in the reverse bias condition to be good. If a silicon diode reads 0.6 to 0.7 in *both* forward and reverse bias, it is bad (shorted) and must be replaced. Also, if a silicon diode reads infinite in *both* the forward and reverse bias conditions, it is bad (open) and must be replaced.

If your DMM does *not* have a diode setting, you can still check a PN junction diode, but you have to use a resistance setting. Using the resistance setting of a DMM is less accurate than using the diode setting. When using the diode setting, the supplied battery voltage actually turns on the diode, and the measured result is the voltage dropped across the junction: @ 0.3 V for a germanium diode and @ 0.6 V for a silicon diode. When using a resistance setting, the supplied battery voltage isn't enough to turn on the diode, so the measured result is the resistance of the diode's junction.

To use a resistance setting of your DMM to test a diode, set your meter to a low setting (1 kΩ or below). Then touch or clip the red (positive) lead of the meter to the anode end of the diode, and touch or clip the black (negative) lead of the meter to the cathode (striped) end of the diode. The meter should display a low resistance (@ 0 Ω to 900 Ω) for a good diode whether it's a germanium diode or a silicon diode. Again, this is called the forward bias condition—positive lead of meter to positive (anode) end of diode and negative lead of meter to negative (cathode) end—which breaks down the barrier or depletion region and causes current to flow through the diode.

Now for the second check. Touch or clip the red (positive) lead of the meter to the cathode (striped) end of the diode, and touch or clip the black (negative) lead of the meter to the anode of the diode. The meter should display a very high resistance or infinite resistance, which may be represented by the infinity symbol "∞" or "OL" (overload) for a good diode whether it's germanium or silicon. Again, this is called the reverse bias condition—positive lead of meter to negative (cathode) end of diode and negative lead of meter to positive (anode) end—which enlarges the depletion region (the "no electron zone") in the diode and prevents the flow of current through the diode.

Consequently, a germanium diode or silicon diode must read a low resistance (usually @ 0 to 900 Ω) in the forward bias condition *and* also read a very high resistance or infinite resistance in the reverse bias condition to be good. If a diode reads a low resistance in *both* the forward and reverse bias conditions, it is bad (shorted) and must be replaced. Also, if a diode reads a very high or infinite resistance in *both* the forward and reverse bias conditions, it is bad (open) and must be replaced.

Whether you use the diode setting or the resistance setting of a DMM to test a diode, you are really checking to see if the junction or barrier or depletion region is intact between the P-type material and the N-type material of a diode. When a diode goes bad due to an open or a short, the junction is broken down, and the diode no longer acts like a semiconductor; that is, it may conduct in both directions or not conduct at all.

Troubleshooting Tip

If you have a diode setting on your DMM, you *must* use it to test diodes. The resistance settings of your DMM might not supply enough voltage to turn on a diode, so you will get false readings.

Checking Diodes with a DMM

Materials, Equipment, and Parts:

- Digital Multimeter (DMM) with test leads.

- Assortment of silicon diodes such as 1N4001, 1N4002, 1N4004, and 1N4007.

- Assortment of germanium diodes such as 1N34, 1N60, and 1N277.

- Assortment of fast-switching diodes such as 1N914 and 1N4148.

Discussion Summary:

A good diode should measure @ 0.6 V for silicon or @ 0.3 V for germanium in the forward bias condition (positive lead of meter to anode of diode and negative lead of meter to cathode of diode) *and* @ infinity (∞) for the reverse bias condition (positive lead of meter to cathode of diode and negative lead of meter to anode of diode). *Note*: a DMM may show OL, which means overload or infinity.

Procedure:

1 Measure and record the barrier voltage (V_B) of each diode below using the diode setting of the DMM, and decide if the diode is good or bad. The first one has been completed to serve as an example.

Diode #	V_B Forward	V_B Reverse	Si/Ge	Good/Bad
1N4001	0.631 V	infinite	Si	Good
_____	_____	_____	_____	_____
_____	_____	_____	_____	_____
_____	_____	_____	_____	_____
_____	_____	_____	_____	_____
_____	_____	_____	_____	_____

NOTE: Complete this section *only* if your multimeter does *not* have a diode setting. Some DMMs don't produce enough voltage on the resistance setting to make a diode conduct in the forward bias condition, so you may get a false reading—that is, you may think a diode is bad when it is really good. If the forward bias measurement is 100 Ω to 900 Ω *and* the reverse bias measurement is very high or infinite, the diode is good.

(continues)

LAB ACTIVITY 1-1

(continued)

2 Measure and record the resistance of each diode using an ohmmeter set on the 2 kΩ setting of your meter (or 1 kΩ setting, depending on what type of meter you have) and decide if the diode is good or bad. The first one has been completed to serve as an example.

Diode #	V_B Forward	V_B Reverse	Si/Ge	Good/Bad
1N60	320 Ω	infinite	Ge	Good
_____	_____	_____	_____	_____
_____	_____	_____	_____	_____
_____	_____	_____	_____	_____
_____	_____	_____	_____	_____
_____	_____	_____	_____	_____

Using Solid State Challenges™ to Test Diodes

Using the PC at your workstation, open the Solid State Challenges™ to the lesson "Basic Diode Testing." Read the introductory screen "Testing Diodes Using an Ohmmeter" and then click on the Begin tab. Read the directions in the red box, and then practice checking diodes until you complete all 11 exercises.

1-4 IDENTIFYING DIODES USING A SEMICONDUCTOR REPLACEMENT GUIDE

Once you've determined a diode is defective, you need to find an equivalent replacement diode. Any diode just won't do! Of course, if you have a bad 1N4001 diode, you can replace it with another 1N4001 diode from your parts bin. If you don't have a 1N4001 diode on hand, then you have to try another approach. Since diodes come in different sizes and have different operating characteristics, you need to find the correct match for the bad diode. This is where semiconductor replacement guides or manufacturer specification (spec) sheets come in handy.

While common devices used in DC and AC circuits—resistors, coils, and capacitors—have their values stamped on them or can be identified by a color code scheme, most semiconductor devices have a manufacturer's part number stamped on their casing. This manufacturer code is needed to identify the type of device—diode, transistor, SCR, etc.—and its operating characteristics. Sometimes the part number of a diode is so small you need a magnifying glass to see it. Other times the part number is faded beyond recognition, is charred from burning out, or isn't listed on a schematic. Don't panic, but keep in mind that you can't replace a bad diode with just any diode in your parts bin. The three most important diode characteristics are size (which is related to power-handling capability), maximum peak reverse voltage (the maximum voltage the

diode can handle in reverse bias), and maximum average forward current (the current the diode can handle in forward bias). Each of these must be considered before replacing a diode with another diode.

A manufacturer's data, or specification (spec), sheet lists everything about a particular diode. Figure 1-5 shows a spec sheet from Fairchild Semiconductor® for a 1N4001 diode. (Actually, the spec sheet lists the characteristics for several diodes, 1N4001 through 1N4007; sometimes diode manufacturers will group similar semiconductors on one spec sheet.) Many semiconductors have multiple spec sheets. For example, only one of the three spec sheets for the 1N4001 diode is shown in Figure 1-5.

The spec sheet shows the diode's features, maximum ratings, thermal characteristics, and electrical characteristics. To replace a diode, however, you only need certain information. The replacement diode should be the same size as the diode you are replacing, so it can handle the circuit power—*never replace a diode with a smaller diode*. From the data sheet, we can see that the maximum peak reverse voltage (or V_{RRM}) for a 1N4001 diode is 50 volts and that the maximum average forward current (I_O or I_F) is 1.0 A (amp).

Naturally, if you have a 1N4001 diode in your parts bin, you can use it to replace a bad 1N4001 diode. However, if you don't have any 1N4001 diodes in your parts bin, then you can use another diode as long as the replacement diode is the same size or bigger than the bad 1N4001

FAIRCHILD SEMICONDUCTOR®

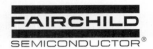

May 2009

1N4001–1N4007 General Purpose Rectifiers

Features

- Low forward voltage drop.
- High surge current capability.

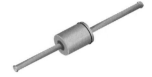

DO-41
COLOR BAND DENOTES CATHODE

Absolute Maximum Ratings* T_A = 25°C unless otherwise noted

Symbol	Parameter	Value							Units
		4001	4002	4003	4004	4005	4006	4007	
V_{RRM}	Peak Repetitive Reverse Voltage	50	100	200	400	600	800	1000	V
$I_{F(AV)}$	Average Rectified Forward Current .375" lead length @ T_A = 75°C	1.0							A
I_{FSM}	Non-Repetitive Peak Forward Surge Current 8.3 ms Single Half-Sine-Wave	30							A
I^2t	Rating for Fusing (t<8.3 ms)	3.7							A^2sec
T_{STG}	Storage Temperature Range	–55 to +175							°C
T_J	Operating Junction Temperature	–55 to +175							°C

* These ratings are limiting values above which the serviceability of any semiconductor device may by impaired.

Thermal Characteristics

Symbol	Parameter	Value	Units
P_D	Power Dissipation	3.0	W
$R_{\theta JA}$	Thermal Resistance, Junction to Ambient	50	°C/W

Electrical Characteristics T_A = 25°C unless otherwise noted

Symbol	Parameter	Value	Units
V_F	Forward Voltage @ 1.0 A	1.1	V
I_{rr}	Maximum Full Load Reverse Current, Full Cycle T_A = 75°C	30	μA
I_R	Reverse Current @ Rated V_R T_A = 25°C T_A = 100°C	5.0 50	μA μA
C_T	Total Capacitance V_R = 4.0 V, f = 1.0 MHz	15	pF

1N4001–1N4007 Rev. C2 www.fairchildsemi.com

FIGURE 1-5 Specification sheet for a 1N4001 diode (Courtesy of Fairchild Semiconductor®)

diode *and* the replacement diode can handle a maximum peak reverse voltage of at least 50 volts *and* the replacement diode can handle an average forward current of at least 1 amp.

Another way to find a replacement diode (or any semiconductor device, for that matter) is to use a semiconductor replacement guide. NTE Electronics, Inc. has a portable *NTE Semiconductor Technical Guide and Cross Reference* catalog that lists over 4,700 equivalent replacement semiconductor devices manufactured by NTE. (The 14th edition, published in 2008, is the latest catalog.) NTE also has QUICKCross™ software, which is available as a free download, as well as an on-line search site. The *NTE* catalog is not as extensive as a manufacturer data or spec sheet, but it lists the critical information for replacing semiconductors: the case style or pin configuration, which identifies the leads, such as the anode and cathode of a diode; the maximum peak reverse voltage; and the maximum average forward current. So, if you have a bad diode, you look up the part number of the diode in the *NTE* catalog to find an equivalent replacement part.

It takes a bit of practice to become proficient in finding replacement parts for semiconductors. The following example uses page numbers and information from the 14th edition of the *NTE* catalog; your edition may have different page numbers. It's important to understand that the procedure is the same for any edition of the *NTE* catalog: start at the back of the book, go to the front of the book, and then go to the middle of the book: back, front, middle. In Semiconductor class, I always tell my students, *"Look it up to hook it up."*

For example, say you have a bad diode with a part number 1N4001. You turn to the *back* of the *NTE* catalog and search for the part number 1N4001. (The back section of the catalog lists part numbers in alphanumeric order; that is, by letters and then numbers.) You will find that diode 1N4001 has an NTE replacement number of 116. Now, you go to the *front* of the *NTE* catalog and look through the numerical index until you find the NTE type number 116, which is on page VII of the 14th edition. On the same line is additional information for the diode: the page number, 1-75; the diagram number, 92; and description. If you turn to page 1-75 and look under NTE type number 116, you will see columns of information. This information includes the material the diode is made of (silicon, or Si); the description and application (general-purpose rectifier); maximum peak reverse voltage (600 V); and the maximum average forward current (1 A). If you look further at diagram 92 on page 1-81, you will see a drawing of the diode with dimensions and the statement *Color Band Denotes Cathode*. Thus, you can use any diode with an NTE replacement number of 116 to replace your bad 1N4001 diode.

INTERNET ALERT

To find semiconductor replacement parts on-line, see www.nteinc.com

If your class does *not* have any *NTE Semiconductor Technical Guide and Cross Reference* catalogs, then you can use the on-line version of the *NTE* catalog. Log on to www.nteinc.com and go to the right panel, Search Central. Click on the button Component Cross Reference, which will bring you to the Cross Reference Search screen. In the text box to the left of the Search button, type in the part number of the bad diode, for example 1N4001, and then click the Search button. The database will respond with the number of matches for 1N4001. In my search, it gave 11 matches, all with an NTE number of 116. I went to the seventh match and clicked on the PDF button under the heading Data Sheets. This opened an NTE116 data sheet, which has information similar to the hard copy of the *NTE* catalog. Of course, there are hundreds of semiconductor parts manufacturers and just as many web sites. We use hard copies of the *NTE* catalog in the classroom because they are tried and true. Again, finding replacement semiconductor parts quickly is an acquired skill that requires a lot of practice.

LAB ACTIVITY 1-3

Using a Semiconductor Guide to Find Replacement Diodes

Materials, Equipment, and Parts:

- *NTE Semiconductor Technical Guide and Cross Reference* catalog

  or

  NTE QUICKCross™ software

  or

  Internet access, www.nteinc.com

- Assortment of rectifier diodes such as 1N4001, 1N4002, 1N4004, and 1N4007.

- Assortment of germanium diodes such as 1N34, 1N60, and 1N277.

- Assortment of fast-switching diodes such as 1N914 and 1N4148.

- Assortment of Zener diodes such as 1N4728, 1N4743, and 1N5248.

Discussion Summary:

Semiconductors such as diodes or transistors can go bad (become shorted or open). Since there are many manufacturers of these semiconductors, a technician needs to be able to identify the part number of a diode or transistor and find an equivalent replacement. NTE Electronics, Inc. is a leading supplier of generic semiconductor replacement parts.

Procedure:

Using one (or all three) of the sources listed above and the part number for each diode, find the diode's NTE replacement number, the composition of the diode (silicon, Si, or germanium, Ge), and a description of the diode. The first two have been completed to serve as examples. *Note:* For Zener diodes, include rated voltage V_Z and wattage under the column for "Description." All Zener diodes are made from a silicon base.

(continues)

LAB ACTIVITY 1-3

(continued)

Diode Part #	NTE #	Si/Ge	Description
1) 1N4001	116	Si	General-purpose rectifier
2) 1N5235B	5014A	Si	Zener diode, 6.8 V, ½ W
3) _____	_____	_____	_____
4) _____	_____	_____	_____
5) _____	_____	_____	_____
6) _____	_____	_____	_____
7) _____	_____	_____	_____
8) _____	_____	_____	_____
9) _____	_____	_____	_____
10) _____	_____	_____	_____
11) _____	_____	_____	_____
12) _____	_____	_____	_____

1-5 USING DIODES IN A DC CIRCUIT

Now that we can identify and test diodes, it's important to see the way diodes operate in a real circuit. During Lab Activity 1-3, we found that there are many different diode descriptions (types) and applications. A PN junction diode is most commonly used as a **rectifier**—to change an AC signal to a DC signal. Other diodes perform as fast-acting switches, some act as light sources, and others provide voltage regulation. These diodes and their applications will be discussed in detail in Chapter 2. For now, we need to understand the behavior of a diode in a simple DC series circuit.

Figure 1-6 shows a circuit diagram with schematic symbols for a 12 V DC source, a 1 kΩ fixed resistor, and a 1N4001 diode, which is made of silicon. Notice that the diode is labeled D_1. The schematic designator for all diodes begins with the letter *D* followed by a number that is usually in subscript form. For example, if you have three diodes in a circuit, they would be labeled D_1, D_2, and D_3.

If you're ever asked to analyze a circuit with a diode in it, *assume the diode is silicon* unless it is specifically labeled germanium. Circuit analyses use a standard 0.3 V drop for germanium diodes in the forward bias condition and a 0.6 V drop for silicon diodes in the forward bias condition. Although one end of the diode is separated from the 12 V source by the 1 kΩ load resistor, you can trace the line and see that the anode of the diode connects to the positive side of the 12 V source, and the cathode of the diode connects to the negative side of the 12 V source. Again, this is the forward bias condition, and the diode will drop about 0.6 V in this case. If you connect the positive lead of a voltmeter to the anode of the diode, and the negative lead of the voltmeter to the cathode of the diode, you will get a reading of @ 0.6 V.

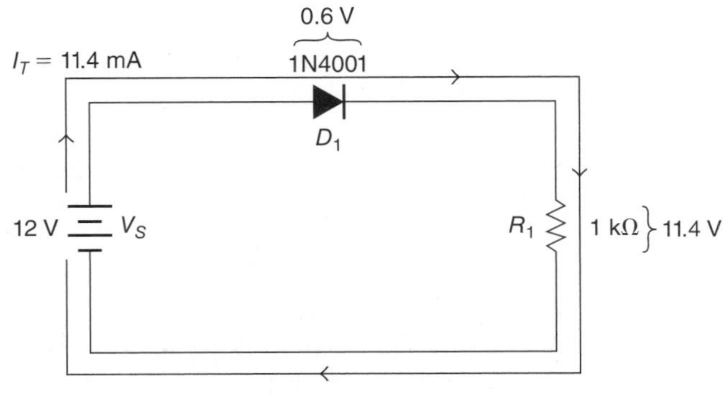

Two important points must be kept in mind when building or testing a diode in a live circuit:

1) *A current-limiting resistor must be put in series with the diode(s) to prevent excessive current from destroying the diode(s).* This resistor is often designated R_S and usually has a value from @ 270 Ω to 1 kΩ.

2) *In order to calculate the total current (I_T) in a diode circuit, go to the resistor.* This is because when the diode is forward biased and is conducting, it acts like a wire that has almost 0 Ω resistance. We can't use Ohm's Law, $I = \frac{V}{R}$, to figure the current through the diode because division by 0 is not defined.

Returning to the circuit of Figure 1-6, we see that we have a DC series circuit. In a series circuit, remember that the voltage drops equal the total, or source, voltage (V_S). If the diode drops @ 0.6 V (V_D) and the total circuit voltage is 12 V, then we can do a little subtraction to determine the resistor voltage (V_R):

EQUATION 1-1

Resistor voltage in series with a diode

$$V_R = V_S - V_D$$
V_R = resistor voltage
V_S = source voltage
V_D = voltage drop across diode

So, for Figure 1-6, $V_R = 12\ V - 0.6\ V = 11.4\ V$

Link to Prior Learning

Ohm's Law

$$I = \frac{V}{R}$$

We can now use the voltage drop across the resistor, the value of the resistor, and Ohm's Law to find the total circuit current (I_T).

So for Figure 1-6, $I_T = \dfrac{11.4\ V}{1\ k\Omega} = 0.0114\ A = 11.4\ mA$.

$I_T = 11.4$ mA
0.6 V
1N4001
D_1
12 V — V_S
R_1 1 kΩ 11.4 V

FIGURE 1-6 Diode in a series DC circuit
(© Cengage Learning 2012)

EQUATION 1-2

Total current of circuit with a diode and resistor

$$I_T = \frac{V_R}{R}$$

I_T = total current

V_R = voltage across resistor

R = resistor value

Let's do some examples to reinforce the calculations we've just learned.

EXAMPLE 1

Situation

Determine V_{R1} and I_T for the circuit of Figure 1-7.

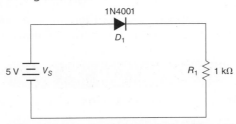

FIGURE 1-7   Diode in a series DC circuit (© Cengage Learning 2012)

Solution

For Figure 1-7,

$$V_R = 5\,V - 0.6\,V = 4.4\,V$$

$$So,\ I_T = \frac{V_R}{R} = \frac{4.4\,V}{1\,k\Omega} = .0044\,A$$

$$= 4.4\,mA$$

Link to Prior Learning

Open Circuit

When a component is open in a series DC circuit, the entire source voltage appears across the open component.

EXAMPLE 2

Situation

Determine V_{R1} and I_T for the circuit of Figure 1-8.

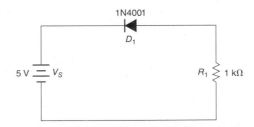

FIGURE 1-8   Diode in a series DC circuit (© Cengage Learning 2012)

Solution

In Figure 1-8, the diode is in reverse bias. When the diode is reverse biased, the depletion region is maximum, the diode acts like an open device, and no current flows through the diode or the rest of the circuit. The entire DC source voltage appears across the diode, so no voltage appears across the resistor.

For Figure 1-8,

$$V_D = 5\,V$$

$$V_R = 0\,V$$

$$I_T = \frac{V_R}{R} = \frac{0\,V}{1\,k\Omega} = 0\,A$$

Of course, we can verify these calculations by analyzing a diode's behavior through circuit simulation software. Lab Activity 1-4 takes you through the process of building a diode circuit on Multisim® 7. No matter what type of software is in your classroom—Multisim®, Electronics Workbench®, or SPICE—the procedure is basically the same: locate the parts and measuring instruments needed from the parts bins; drag them onto the "workbench" or desktop; change the values if needed; connect all parts and instruments together; and then power on and verify results.

LAB ACTIVITY 1-4

Constructing a Diode Circuit Using Multisim® 7

Materials, Equipment, and Parts:

- PC w/Multisim® 7

Procedure:

1 Open Multisim® 7 to the desktop screen.

2 In the left panel, click on the icon Show Power Source Components Bar (the icon looks like the schematic symbol for a DC voltage source), and then double-click the icon Place DC Voltage Source. The schematic symbol for a DC voltage source will appear on the desktop. Close the Power Source Component box.

3 Use your mouse to click and drag the schematic symbol for a DC voltage source to column 0, row C (the left center of the desktop).

4 Double-click the schematic symbol and change the voltage to 5 V. Click OK.

5 In the left panel, click on the Show Basic Components Bar (the icon looks like the schematic symbol for a fixed resistor), and then double-click the icon Place Resistor. The schematic symbol for a resistor will appear on the desktop. Close the Basic Components window.

6 Use your mouse to click and drag the resistor symbol to column 3, row B. (If the resistor value is not 1 kΩ, double-click the resistor schematic symbol and change the value to 1 kΩ.)

7 In the left panel, click on the gray icon diode (the icon looks like the schematic symbol for a diode). The Select a Component dialog box opens. In the left panel, click on DIODE and then click on 1N4001GP. Click OK. The diode labeled 1N4001GP should appear on the desktop.

8 Use your mouse to click and drag the diode symbol to column 4, row C.

(continues)

LAB ACTIVITY 1-4

(continued)

9   Right-click the diode and select 90 clockwise. The diode will rotate to a position with the arrow pointing down.

10   In the left panel, click on the icon Indicator (the icon looks like a red 8 within a box). The Select a Component dialog box opens. In the left panel under Database, click on AMMETER. In the middle panel Component, click on AMMETER_H. Click OK. The ammeter should appear on the desktop.

11   Click and drag the ammeter to column 1, row B.

12   In the left panel, click on the icon Source (the icon looks like the schematic symbol for ground). The Select a Component dialog box opens. In the left panel under Database, click on POWER_SOURCE. In the middle panel under Component, click on GROUND. Click OK. The ground symbol should appear on the desktop.

13   Click and drag the ground symbol to column 2, row D.

14   Connect all the components together in series with each other.

15   Click on the Run/Stop simulation switch in the upper right side of the screen. This applies power to the circuit.

After applying simulated power to the circuit, what is the value of total current?

$I_T =$ _____

How does this value compare to the calculated I_T of Figure 1-7?

Go back to the icon Indicator, select a voltmeter, and place it on the desktop. Hook it to the circuit to measure and record the following values:

V_D _____

V_R _____

How do these voltage values compare to the calculated values of Figure 1-7?

Constructing a Diode Circuit Using a Breadboard

Materials, Equipment, and Parts:

- Digital Multimeter (DMM) with test leads.
- 5 V DC power source.
- Breadboard and connecting wires.
- 1N 4001 or equivalent silicon diode.
- 1 kΩ fixed resistor.

Discussion Summary:

In a series circuit containing a DC voltage source, silicon diode, and resistor, the conducting diode will drop @ 0.6 volts. The remainder of the DC source voltage will appear across the resistor. The total circuit current (I_T) can be found using the following equation:

$$I_T = \frac{V_R}{R}$$

Procedure:

SAFETY FIRST. Eye protection should always be worn when working with live voltages. Before powering on a live circuit, always check with your instructor.

1 Set your DMM to measure DC voltage. Measure and then record the V_{OUT} of your DC power supply. If it is a variable supply, you may have to adjust your output until it reads @ 5 V on your DMM.

DC power supply measured voltage: _____

2 Power off. Measure and record the value of your 1kΩ resistor.

Measured resistor value: _____

3 On the breadboard, construct the circuit of Figure 1-9. Have your instructor check the circuit.

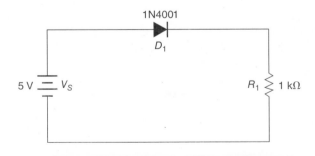

FIGURE 1-9   Diode in a series DC circuit
(© Cengage Learning 2012)

(continues)

LAB ACTIVITY 1-5

(continued)

4 Power on and then use your DMM to measure and record the following values:

V_D _____

V_R _____

5 Compare your measured values to the calculated values of Figure 1-7.

6 Turn off circuit power. Break open the circuit and insert your DMM in series with the resistor. Switch your meter setting to measure DC current. Power on and then record the I_T for the circuit.

I_T _____

7 Compare your measured current to the calculated value of Figure 1-7.

8 Explain any noticeable differences between the calculated values of Figure 1-7 and the values you just measured.

1-6 THE DIODE AS A RECTIFIER IN AN AC CIRCUIT

Earlier in this chapter, we mentioned that PN junction diodes are most commonly used as rectifiers—to change an AC signal to a DC signal. Chapter 3 goes deeply into the subject of rectifiers and how they are used in power supplies. For now, we'll take a look at the basic process of rectification.

Figure 1-10(a) shows a series circuit with a 6.3 V AC source, a 1N4001 or equivalent silicon diode, and a 12 kΩ fixed resistor. The 12 kΩ resistor is called the load resistor and is labeled R_L. Remember from DC that a load resistor consumes the power in any electronic circuit. Examples of loads are computers, toasters, and TVs. Since the 12 kΩ resistor is in series with the diode, it also acts as a current-limiting resistor.

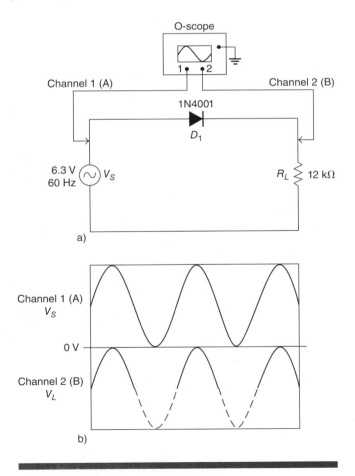

a)

b)

FIGURE 1-10 Diode as rectifier in a series AC circuit: a) Schematic and b) O-scope readings (© Cengage Learning 2012)

The 6.3 V AC source is on the left, the anode of the diode connects to the top side of the source, the cathode of the diode connects to the top of the 12 kΩ load resistor, and the bottom of the 12 kΩ load fixed resistor connects to the bottom side of the 6.3 V AC source. A dual trace oscilloscope is connected to the circuit. Channel 1, or A, of the oscilloscope is connected to the 6.3 V AC source and shows an AC sine wave on the oscilloscope screen (the waveform on the top). Channel 2, or B, of the oscilloscope is connected across the 12 kΩ load resistor.

As you can see from the oscilloscope display in Figure 1-10(b), the AC sine wave of Channel 2, or B, (the waveform on the bottom) is missing its negative alternation. This is because the diode passes current only on the positive alternation; that is, when the diode is forward biased. When the 6.3 V AC source changes direction on the negative alternation, the diode is reverse biased and does not conduct. Thus, the waveform taken across the 12 kΩ load resistor shows a blue dotted line where the negative alternation should be. The diode has provided rectification; it has changed the AC signal to a pulsating DC signal. In Chapter 3, we'll explore this phenomenon in detail.

CHAPTER SUMMARY

Since the late 1950s, solid-state devices or semiconductors have replaced vacuum tubes in radios, TVs, computers, and industrial electronics. Semiconductors have several advantages over vacuum tubes: they use less power, are more rugged, last longer, and take up less space. Vacuum tubes are still used today in limited applications such as very high-power radio and TV transmitters and in audio amplifiers used by musicians.

Semiconductors are composed of either a silicon base or a germanium base. Most semiconductors use a silicon base because silicon is cheaper to manufacture and because it withstands heat far better than germanium. The PN junction diode is the simplest and most commonly used semiconductor device; it is made of P-type material (such as boron, which has a shortage of electrons) and N-type material (such as arsenic, which has a surplus of electrons) that are separated by a barrier or depletion region. For a good germanium diode, the voltage between the P-type material and the N-type material will measure @ 0.3 volts; for silicon, the voltage will measure @ 0.6 volts. The end of the diode connected to the P-type material is called the anode, and the end of the diode connected to the N-type material is called the cathode, which often has a stripe for identification purposes. The cathode of a diode is usually connected to the negative side or ground in a circuit.

Electronic technicians use the diode setting of a digital multimeter (DMM) to measure diodes and other semiconductor devices. When checked with the diode setting, the diode is forward biased (positive lead of DMM to anode of diode and negative lead of DMM to cathode), the barrier or depletion region breaks down, and current flows through the diode. In the reverse bias condition (positive lead of the meter to the cathode end of the diode and negative lead of the meter to the anode of the diode), the depletion region is increased, no current flows, and the meter display should show infinite.

Semiconductor replacement guides or manufacturer specification (spec) sheets are used to find equivalent replacements for bad diodes. These guides list the critical information for replacing semiconductors: the case style or pin configuration, which identifies the leads, such as the anode and cathode of a diode; the maximum peak reverse voltage; and the maximum average forward current.

PN junction diodes are most commonly used as rectifiers—to change an AC signal to a DC signal. Other diodes perform as fast-acting switches, and some provide voltage regulation.

CHAPTER EQUATIONS

Resistor voltage in series with a diode

(Equation 1-1)

$$V_R = V_S - V_D$$

V_R = resistor voltage

V_S = source voltage

V_D = voltage drop across diode

Total current of circuit with a diode and resistor

(Equation 1-2)

$$I_T = \frac{V_R}{R}$$

I_T = total current

V_R = voltage across resistor

R = resistor value

CHAPTER REVIEW QUESTIONS

Chapter 1-1

1. List three advantages that solid-state devices (semiconductors) have over vacuum tube devices.

2. Why haven't semiconductors replaced vacuum tubes in all electronic applications?

Chapter 1-2

3. How much external voltage must be applied across a germanium diode to overcome the barrier voltage and cause the diode to conduct?

4. How much external voltage must be applied across a silicon diode to overcome the barrier voltage and cause the diode to conduct?

5. What alphanumeric prefix is used for the part number of a diode?

6. How can one tell the cathode of a diode just by looking at it?

7. Draw and label the schematic symbol for a diode.

8. Why are most diodes made of silicon rather than germanium?

Chapter 1-3

9. What is bias?

10. Explain the difference between forward bias and reverse bias.

11. When does a diode allow current to pass through it?

12. What will a good germanium or silicon diode measure in reverse bias?

13. If a silicon diode reads @ 0.6 V in both forward and reverse bias, is it good or bad?

14. In the forward bias condition, will a good silicon diode have low resistance or high resistance?

15. When using the diode setting of a DMM to test a diode, what are you actually checking?

Chapter 1-4

16. When looking for a replacement diode, what are the three most important characteristics to consider?

17. What is the maximum peak reverse voltage rating of a diode?

18. Why shouldn't you replace a diode with a smaller diode?

19. Use an *NTE* catalog, manufacturer data sheet, or the Internet to find the NTE replacement part for a diode with the part number 1N4733.

Chapter 1-5

20. What letter is used as the schematic designator for diodes?

21. In a live circuit, why must any diode have a resistor in series with it?

22. A forward-biased silicon diode is in series with a 12 V DC source and a 560 Ω resistor. What is the voltage across the resistor (V_R)?

23. What is the total current (I_T) of the circuit in Question 22?

24. **What is the voltage across the diode in Figure 1-11? Why?**

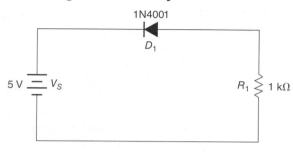

FIGURE 1-11   Diode in a series DC circuit (© Cengage Learning 2012)

Chapter 1-6

25. **A PN junction diode is commonly used as a rectifier in an AC circuit. What does a rectifier do?**

Special-Purpose Diodes and Their Applications

O B J E C T I V E S   *Upon completion of this chapter, you should be able to:*

- Draw and label the schematic symbol for a light-emitting diode (LED).
- Explain the operation of an LED.
- Measure an LED with a digital multimeter and determine if it is good or bad.
- Construct a circuit with a common-anode seven-segment LED display and explain its operation.
- Draw and label the schematic symbol for a Zener diode.
- Measure a Zener diode with a digital multimeter and determine if it is good or bad.
- Explain the operation and application of Zener diodes.
- Construct a voltage regulation circuit using a Zener diode.

- Draw and label the schematic symbol for a photodiode.
- Explain the role of a photodiode in an optoisolator circuit.
- Construct an optoisolator integrated circuit.
- Draw and label the schematic symbol for a varactor diode.
- Explain the operation of a varactor diode and its most common application.
- Draw and label the schematic symbol for a Schottky diode and explain its operation.
- Draw and label the schematic symbol for a tunnel diode and explain its operation.
- Draw and label the schematic symbol for a PIN diode and explain its operation.

MATERIALS, EQUIPMENT, AND PARTS

Materials, equipment, and parts needed for the lab experiments in this chapter are listed below:

- *NTE* catalog or Internet access, www.nteinc.com
- Digital Multimeter (DMM) with test leads.
- 5 V DC voltage source.
- 12 V DC voltage source.
- Variable DC voltage source, 0 V to 15 V.

- Breadboard and connecting wires.
- Two-terminal LED (any color).
- Common-anode seven-segment single-digit LED display, MAN4610A or equivalent.
- Zener diode with a V_Z of 5 V to 12 V. Some common Zener diodes within this range are

1N4733A (5.1 V), 1N5235B (6.8 V), and 1N5240B (10 V).
- 4N25 or 4N35 optoisolator IC.
- SPST switch.
- 270 Ω fixed resistor (4), 820 Ω fixed resistor, 1 kΩ fixed resistor, and 39 kΩ fixed resistor.

GLOSSARY OF TERMS

Light-emitting diode (LED) A semiconductor device that gives off light when a sufficient voltage is applied to it

Seven-segment LED display A semiconductor device that contains seven or more LEDs constructed in a sealed unit and is used to display numbers and letters in applications such as calculators and automobile instrument panels

Common-anode seven-segment display A seven-segment LED display constructed so the anodes of all the LEDs are linked internally and then connected to two external pins. The cathode of each internal LED is connected to an external pin. To light a particular LED segment, 5 V DC is applied to one of the common-anode pins, and the cathode pin of the particular LED is connected to ground through an external resistor

Common-cathode seven-segment display A seven-segment LED display constructed so the cathodes of all the LEDs are linked internally and then connected to two external pins. The anode of each internal LED is connected to an

external pin. To light a particular LED segment, one of the two common-cathode pins is connected to ground through an external resistor, and 5 V DC is applied to the anode pin of the particular LED

Organic light-emitting diode (OLED) A light-emitting diode made of carbon atoms encased in metal anodes and cathodes, which in turn are encased in glass plates; it is extremely thin, flexible, and has low power consumption

Zener diode A semiconductor device that conducts when a reverse bias voltage applied to its terminals exceeds a certain value called the V_Z; once conducting, a Zener maintains a steady voltage across its terminals, making it useful as a voltage regulator in power supplies

Photodiode A diode that produces a voltage when struck by light; it is used in solar cells and solar panels

Optoisolator A semiconductor device made of an LED-to-photodiode combination that uses light to join an external source to an external load; it is also called an optocoupler

Integrated circuit An electronic device that contains several electronic components housed in one package, often on a semiconductor wafer or chip

Varactor A semiconductor device that varies its capacitance according to an applied reverse bias voltage and is used in automatic tuning applications

Schottky diode A semiconductor device that has a low turn on/turn off voltage, making it useful as a very high-speed electronic switch; it is also known as a hot-carrier diode

Tunnel diode A semiconductor device that provides high-speed switching, making it useful in communication systems, particularly high-frequency applications such as local oscillators for UHF TV tuners; it is also known as the Esaki diode

PIN diode A specialized diode that behaves like a varactor when reverse biased and like an RF (radio frequency) switch when forward biased, making it useful in high-frequency test probes and in radio communication applications

2-1 LIGHT-EMITTING DIODES (LEDs) AND SEVEN-SEGMENT LED DISPLAYS

In Chapter 1, we covered in depth the PN junction diode. Although the PN junction diode can be found in nearly every electronic circuit, it is not easy to spot unless you've had some training in electronics. The **light-emitting diode (LED)**, on the other hand, is clearly visible to anyone and is found in everything from panel indicator lights to clocks to digital billboards to headlights on newer cars and trucks. First marketed in the 1960s, LEDs soon replaced miniature incandescent (metal filament) light bulbs and lamps in many applications because LEDs were more durable, consumed less power, and lasted longer.

Figure 2-1(a) shows several examples of discrete (stand-alone) two-terminal LEDs. LEDs come in a variety of colors, but only green, yellow, and red are pictured here. If you look closely at the picture, you can see that one terminal, or lead, is longer than the other. The longer lead on an LED is the anode, or positive side, and the shorter lead is the cathode, or negative side. If both leads are the same length, look closely for a slightly flattened edge on the LED dome. The lead closest to this flattened edge is the cathode. As with the PN junction diode, the cathode of an LED is usually connected to the negative side or ground in a circuit.

Figure 2-1(b) shows the schematic symbol for a two-terminal LED. Notice that the symbol for an LED is similar to a PN junction diode except for the addition of two arrows or "lightning bolts" pointing outward. These arrows or lightning bolts are typically used in electronics to represent light. Thus, the LED schematic symbol has light emitting from, or coming out of, the diode. That's why it's called a light-emitting diode.

The operation of an LED is slightly different from that of a PN junction diode. Like the PN junction diode, an LED is made of N-type material (excess electrons) and P-type material (excess holes) plus the chemical element gallium. Unlike the PN junction diode, however, when a forward bias voltage is applied to the LED (again, positive lead of voltage source to anode of LED and negative lead of voltage source to cathode of LED), the interaction between electrons and holes causes the LED to glow. *Put simply, voltage is applied to an LED, and it gives off light.* The color of light emitted by the LED depends on the chemical composition: gallium, aluminum, and arsenic produce red light; gallium and phosphorous produce green light; and other combinations produce yellow, blue, and infrared light. Infrared light is invisible to the naked eye; it is used in many applications such as night-vision goggles, TV remote controls, digital cameras, and fiber optics.

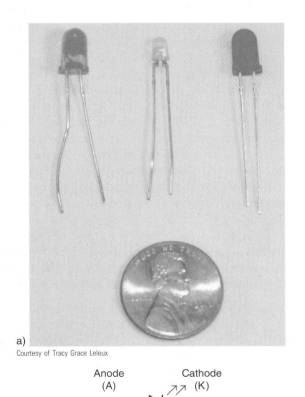

a)

Courtesy of Tracy Grace Leleux

b)

Anode (A)   Cathode (K)

+   −

© Cengage Learning 2012

FIGURE 2-1 Light-emitting diode: a) Discrete two-terminal LEDs and b) Schematic symbol

INTERNET ALERT

To learn more details about the LED, see the website http://electronics.howstuffworks.com/led.htm

LEDs have several advantages over traditional incandescent light bulbs or lamps: they use less power, give off less heat, and are more rugged. In the past, LEDs didn't have the light intensity or power-handling capabilities of standard light bulbs. However, since LEDs first appeared commercially in the 1960s, evolving manufacturing technology is rapidly making LEDs the light source of choice for residential, industrial, and military applications.

Because LEDs are so common in electronics, technicians need to determine if an LED is good or bad. Testing a two-terminal LED is easy. First, set your DMM to the diode setting. Connect the positive lead of your meter to the anode (longer) lead of the LED, and then connect the negative lead of your meter to the cathode of the LED. If both leads are the same length, then look for a flat edge on the LED dome; the lead closest to this flat edge is the *cathode*. The LED should glow. (To see the glow from smaller LEDs, you may have to position the LED so you are looking down at its top.) Your meter should also show a reading of @ 1.5 V to 2 V, the typical voltage drop across an LED in the forward bias condition (V_F). If the LED does not give off light, it is open and needs to be replaced.

Seven or more LEDs can be manufactured in one sealed unit called a **seven-segment LED display**, which is used in calculators, automobile instrument panels, and countless other applications. In a seven-segment display, each segment is an LED. Many seven-segment displays also have an LED for the decimal point. The decimal point can be on the left side of the seven-segment display and is called a left-hand decimal point (LHDP), or leading decimal point. If the decimal point is on the right side of the display, it is called a right-hand decimal point (RHDP), or trailing decimal point. Figure 2-2 shows two common seven-segment displays.

The display on the left in Figure 2-2 is a true seven-segment LED display. Each seven-segment display glows with a particular color light. Common colors used for seven-segment LED displays are red, orange, yellow, and green. The display on the right is a liquid crystal display (LCD), which uses black characters and a gray background often found in calculators and DMMs.

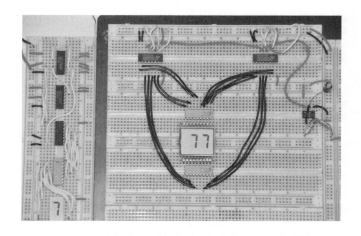

FIGURE 2-2 Seven-segment orange LED and seven-segment liquid crystal display (Courtesy of Tracy Grace Leleux)

LCDs don't use light-emitting diodes. Figure 2-2 clearly shows the difference between LED and LCD displays.

Seven-segment LED displays come in two configurations: common-anode and common-cathode. In the **common-anode seven-segment display**, the anodes of all the LEDs are linked internally and then connected to two external pins; thus, the name common-anode. (One external pin is a spare.) The cathode of each internal LED is connected to an external pin. To light a particular LED segment, 5 V DC is applied to one of the common-anode pins, and the cathode pin of the particular LED is connected to ground through an external resistor. The external resistor, usually 270 Ω to 1 kΩ, limits the current through the LED and prevents it from being burned out. In fact, when one of the segments of a seven-segment display doesn't light, it's usually because its external resistor has burned out (opened). Figure 2-3 shows the internal connections for a common-anode seven-segment display with a right-hand decimal point.

Notice that each LED segment is labeled with a lower-case letter. The top horizontal segment is *a*, and then the labeling scheme continues alphabetically in clockwise order to the segment *f*. The middle segment is labeled *g*. The anodes of all LEDs are connected to both pins 3 and 14. One of these pins will be connected to 5 V DC; the other pin is a spare.

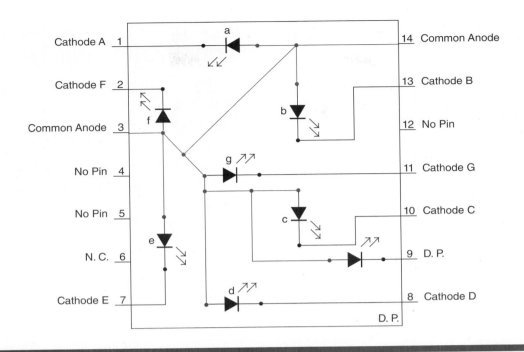

FIGURE 2-3   Internal diagram of common-anode seven-segment display with RHDP (© Cengage Learning 2012)

The cathode for each internal LED is connected to an external pin. For example, cathode *A* is connected to pin 1, cathode *B* to pin 13, etc. Notice that terminals 4, 5, and 12 are labeled "No Pin." The seven-segment display was manufactured this way, so don't panic if you find a seven-segment display that's missing pins. Pin 6 is labeled *N.C.*, which means "no connection" (internally). Sometimes manufacturers leave pins without any electrical function to help stabilize the display on a circuit board. Also, notice that pin 9 is labeled *D.P.*, which stands for "decimal point."

The **common-cathode seven-segment display** is constructed so that the cathodes of all the LEDs are linked internally and then connected to two external pins; thus, the name common-cathode. (One external pin is a spare.) The anode of each internal LED is connected to an external pin. The common-cathode works just the opposite of the common-anode. To light a particular LED segment, one of the two common-cathode pins is connected to ground through an external resistor, and 5 V DC is applied to the anode pin of the particular LED. You can't tell by looking at a seven-segment display if it's common-anode or common-cathode: you need a semiconductor replacement guide, such as the *NTE* catalog.

Whether you use a two-terminal LED or a seven-segment LED display, keep in mind the precaution we learned from Chapter 1: *a current-limiting resistor (R$_S$) must be put in series with a diode to prevent excessive current from destroying the diode.* The value of the R$_S$ depends on the applied source voltage (V$_S$) and the forward current (I$_F$) of the LED; that is, the current through the LED when forward biased. Most LEDs need an I$_F$ of 10 mA to 20 mA to light, with a midrange of 15 mA, and usually drop @ 1.5 V in the forward bias condition (V$_F$). If we have a 5 V DC source, we can use a variation of Ohm's Law to calculate the value of the current-limiting resistor R$_S$ that must be placed in series with the LED.

EQUATION 2-1

Current-limiting resistor

$$R_S = \frac{V_S - V_F}{I_F}$$

V$_S$ = source voltage

V$_F$ = voltage drop across LED in forward bias

I$_F$ = current through LED in forward bias

So, the value of $R_S = \dfrac{5\,V - 1.5\,V}{15\,mA} = 233.33\,\Omega$

A 233 Ω resistor is not a common value, so we'll use the closest value that's higher than 233 Ω, which is 270 Ω. A 330 Ω resistor will also work well. The importance of an external current-limiting resistor can't be stressed enough.

The main disadvantage of a seven-segment display is that if one internal LED burns out, the entire seven-segment display has to be replaced. Lab Activity 2-1 focuses on the two-terminal LED and the common-anode seven-segment LED display.

Troubleshooting Tip

The first step in troubleshooting is to do a visual inspection. If one segment of a seven-segment display doesn't light, first check the current-limiting resistor in series with the LED segment. If the resistor is good, you'll most likely have to replace the entire seven-segment display.

LAB ACTIVITY **2 - 1**

Testing LEDs and Seven-Segment Displays

Materials, Equipment, and Parts:

- *NTE* catalog or Internet access, www.nteinc. com (*Look it up to hook it up.*)

- Digital Multimeter (DMM) with test leads.

- 5 V DC voltage source.

- Breadboard and connecting wires.

- Two-terminal LED (any color).

- Common-anode seven-segment single-digit LED display, MAN4610A or equivalent.

- 270 Ω fixed resistor (4).

Discussion Summary:

A good LED should glow in the forward bias condition (positive lead of meter to anode of diode and negative lead of meter to cathode of diode) and will typically measure a V_F of 1.5 V to 2.0 V across its terminals. In a common-anode seven-segment LED display, one of the common-anode pins is connected to 5 V DC, and the cathode pin of a particular LED segment is connected to ground through an external resistor. Each LED must have its own current-limiting resistor to prevent the LED from burning out.

Procedure:

SAFETY FIRST. Eye protection should always be worn when working with live voltages. Before powering on a live circuit, always check with your instructor.

1 Use your DMM to measure the two-terminal LED at your workstation and verify that it lights. Record the voltage drop in the forward bias condition (V_F).

$V_F =$ _____

2 Using a source voltage V_S of 5 V, the voltage drop V_F of the LED from Step 1 and a current of 15 mA, calculate the value of the current-limiting resistor R_S.

$$R_S = \frac{V_S - V_F}{I_F} =$$

$R_S =$ _____

3 Use this R_S value (270 Ω) to construct the circuit in Figure 2-4.

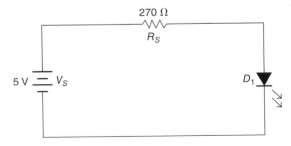

FIGURE 2-4   LED in a series DC circuit
(© Cengage Learning 2012)

(*continues*)

LAB ACTIVITY 2-1

(continued)

4   Power on your DC voltage source and see if the LED lights. Yes _____ No _____
Have the instructor verify your results.

5   Write the part number of the common-anode seven-segment display at your workstation.
Then use the *NTE* catalog, or visit the website www.nteinc.com to find the NTE replacement
number.

Part number _____

NTE number _____

6   Draw and label the pin configuration.

7   At your workstation, connect the +5 V lead of the DC voltage source to one of the red
banks on the breadboard. Connect a wire between the bank and one of the two common-
anode pins on the seven-segment display.

8   Connect the ground lead of the DC voltage source to one of the blue banks on the breadboard. Place the 270 Ω resistor on the breadboard and connect a wire from the blue bank to
one end of the resistor.

9   Connect a wire to the other end of the 270 Ω resistor. Power on the DC voltage source and
use this wire to probe each cathode pin of the seven-segment display. Note which segment
lights for each pin and compare these results to your pin configuration drawing from Step 6.

10   What pins and letters are needed to make the number 3 light?

11   What pins and letters are needed to make the number 7 light?

12   What pins and letters are needed to make the letter *F* light?

13   What pins and letters are needed to make the letter *L* light?

14   Using four 270 Ω fixed resistors, wire the seven-segment display to light the number 4. Have
the instructor verify your results.

Lab Activity 2-1 showed the operation of two-terminal LEDs and seven-segment displays that have been around for several decades. Introduced to the marketplace in the 1990s, the **organic light-emitting diode (OLED)** is a newer LED technology that is gaining ground in electronic applications. An OLED (rhymes with "O-MED") is made of carbon atoms encased in metal anodes and cathodes, which in turn are encased in glass plates. When a DC voltage is applied to the anode and cathode of an OLED, the carbon atoms give off a bright light. Because OLEDs can be made as thin as a hair and have low power consumption, they are being marketed to replace the older heat-intensive picture tube televisions and monitors as well as LCD thin-screen TVs. OLEDs are also being used for displays in cell phones and for commercial and residential lighting.

2-2 Zener Diodes and Voltage Regulation

Diodes are often manufactured for particular applications. The PN junction diode is commonly used for rectification, changing AC to pulsating DC, while LEDs are used as light indicators and other visual displays. The **Zener diode**, named after physicist Clarence Zener, is a diode specially designed for voltage regulation. *The Zener diode's main purpose in a circuit is to protect the expensive load*—that is, whatever is consuming the power in a circuit and performing the most important task. For example, the motor in a washer or dryer is a load; the CPU, monitor, and disk drives in a computer are loads; and the computer module in a vehicle is a load. The Zener diode provides protection for a load through voltage regulation; that is, the Zener diode protects its load from *excessive* voltage.

Figure 2-5(a) shows several examples of Zener diodes. Zener diodes look like PN junction diodes, so the only way to tell if it's a Zener diode is to look up its part number in the *NTE* catalog or another semiconductor replacement guide. *Look it up to hook it up.*

Figure 2-5(b) shows the schematic symbol for a Zener diode. The symbol for a Zener is similar to the symbol for a PN junction diode except that the cathode has two additional lines forming 45° angles. Like the PN junction diode and the LED, the cathode of the Zener is at the pointed end of the triangle.

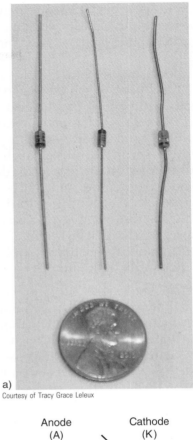

a)
Courtesy of Tracy Grace Leleux

Anode (A)   Cathode (K)

b)
© Cengage Learning 2012

FIGURE 2-5 Zener diode: a) Examples of Zener diodes and b) Schematic symbol

A Zener diode might look like a PN junction diode, but its behavior is drastically different in a circuit. A PN junction will be destroyed if the voltage across its terminals in reverse bias exceeds a certain value known as the V_{RRM} (peak repetitive reverse voltage) or P_{RV} (peak reverse voltage); that is, the diode will break down and then conduct in both directions, so it is no longer a semiconductor.

The Zener diode, however, takes advantage of this situation. The Zener diode is manufactured such that when the reverse bias voltage exceeds a certain value called the V_Z, the Zener starts conducting. The Zener then passes current but maintains a steady voltage across its terminals, and anything connected in parallel (shunt) with the Zener—such as a load—will have the same voltage. Thus, the Zener diode can be used as a voltage regulator, protecting its load from excessive voltage.

You test a Zener diode the same way you test a PN junction diode. First, turn your meter's rotary selection switch to the diode setting. Then touch or clip the red (positive) lead of the meter to the anode end of the diode, and touch or clip the black (negative) lead of the meter to the cathode (striped) end of the diode. Again, this is the forward bias condition, and the meter should display between 0.6 and 0.7 (volt) for a good diode. (Actually, some may read *slightly* less or more.) Next, touch or clip the red (positive) lead of the meter to the cathode (striped) end of the diode, and touch or clip the black (negative) lead of the meter to the anode of the diode. Again, this is the reverse bias condition, and the meter should display infinite, which may be represented by the infinity symbol (∞) or "OL" (overload), for a good diode.

Figure 2-6 shows the operation of a Zener circuit. The source voltage V_S is a variable voltage source with a range of 0 V to 15 V DC. The 1 kΩ resistor R_S limits the current through the Zener, and the 39 kΩ resistor is the load resistor R_L. The Zener diode has a V_Z, or reverse bias breakdown voltage, of 6.8 V. If you look closely, you'll

recognize that Figure 2-6 is a series-parallel circuit: the current-limiting resistor R_S is in series with the parallel combination of the Zener diode and the 39 kΩ load resistor R_L.

Notice that the Zener diode is *purposely hooked up in reverse bias*; the anode end of the Zener is connected to the negative side or ground of the circuit. This is a radical departure from what we've learned about diodes so far. PN junction diodes and LEDs have their anodes connected to the more positive side of a circuit.

In Figure 2-6(a), the source voltage V_S is set to 4 V. Since this value is below the Zener's V_Z, the Zener is not conducting yet; the circuit current goes through the load resistor R_L, and the voltage drop across R_L is nearly the same as V_S. We can use variations of Ohm's Law to determine circuit voltages and currents.

EQUATION 2-2

Zener off

$$I_T = \frac{V_T}{R_T}$$

$$R_T = R_S + R_L$$

So, for Figure 2-6(a), $I_T = \dfrac{V_T}{R_T} = \dfrac{4\ V}{40\ k\Omega} = 0.1\ mA$

Then the voltage across the current-limiting resistor R_S can be determined:

$$V_{RS} = (I_T)(R_S) = (0.1\ mA)(1\ k\Omega) = 0.1\ V$$

Finally, the voltage across the load resistor R_L can be found:

$$V_{RL} = (I_T)(R_L) = 3.9\ V$$

You can see from the equation that the voltage across the load R_L is nearly the same as the 4 V source voltage V_S.

In Figure 2-6(b), the source voltage V_S is now 8 V, which simulates a changing input voltage. The Zener's reverse bias breakdown voltage V_Z of 6.8 V has been exceeded, and the Zener is now conducting, diverting current from the load. The Zener is also maintaining a voltage drop of @ 6.8 V across itself *and* the 39 kΩ load resistor R_L because they are in parallel.

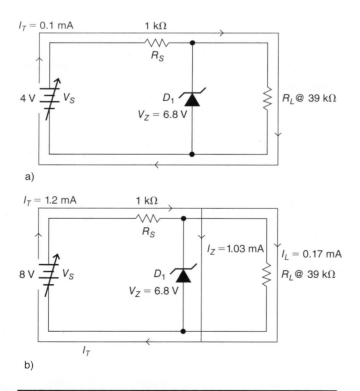

FIGURE 2-6 Operation of Zener diode circuit: a) Zener diode off (not conducting) and b) Zener diode on (conducting) (© Cengage Learning 2012)

The Zener is protecting the R_L from voltage (and thus current) changes that could possibly destroy the load. The circuit voltages and currents have changed from Figure 2-6(a) when the Zener was off.

Zener *on*:

$V_{RL} = V_Z = 6.8$ V (because devices in parallel have the same voltage)

Because V_S, R_S, and the Zener diode form a series circuit, we can rearrange Kirchhoff's Voltage Law to determine V_{RS}:

$V_{RS} = V_S - V_Z = 8$ V $- 6.8$ V $= 1.2$ V

For the total current I_T, we go to the R_S.

EQUATION 2-3

Zener on

$$I_T = \frac{V_{RS}}{R_S}$$

V_{RS} = voltage across current-limiting resistor R_S

So, for Figure 2-6(b), $I_T = \dfrac{1.2 \text{ V}}{1 \text{ k}\Omega} = 1.2$ mA

We can also determine the current through the load resistor R_L:

$$I_{RL} = \frac{V_{RL}}{R_L} = \frac{6.8 \text{ V}}{39 \text{ k}\Omega} = 0.17 \text{ mA}$$

Since the Zener diode and the load resistor R_L are in parallel, we can rearrange Kirchhoff's Current Law to determine the current through the Zener diode.

EQUATION 2-4

Current through a Zener diode

$I_Z = I_T - I_{RL}$
I_Z = current through Zener diode
I_{RL} = current through load resistor

So, for Figure 2-6(b), $I_Z = 1.2$ mA $- 0.17$ mA $= 1.03$ mA.

Looking at the results of Equations 2-3 and 2-4, you can clearly see that the Zener diode has diverted most of the total circuit current away from the load resistor R_L and at the same time provided voltage regulation for the load resistor R_L. The load current is only 0.17 mA, while the current through the Zener diode is 1.03 mA. The Zener has protected the 39 kΩ load from a varying voltage source.

Let's do another example to reinforce the calculations we've just learned.

EXAMPLE 1

Situation

Determine V_{RL}, V_{RS}, I_T, I_{RL}, and I_Z for the circuit of Figure 2-7.

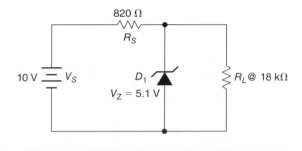

FIGURE 2-7   Zener diode circuit
(© Cengage Learning 2012)

Solution

For Figure 2-7, the Zener diode is on because its reverse bias breakdown voltage V_Z of 5.1 V has been exceeded.

$V_{RL} = V_Z = 5.1$ V   (because devices in parallel have the same voltage)

$V_{RS} = V_S - V_Z = 10$ V $- 5.1$ V $= 4.9$ V

$$I_T = \frac{V_{RS}}{R_S} = \frac{4.9\ V}{820\ \Omega} = 5.97\ mA$$

$$I_{RL} = \frac{V_{RL}}{R_L} = \frac{5.1\ V}{18\ k\Omega} = 0.283\ mA$$

$$I_Z = I_T - I_{RL} = 5.97\ mA - 0.283\ mA$$
$$= 5.68\ mA$$

Again, you can see that the Zener diode has diverted most of the total circuit current away from the load resistor R_L and at the same time provided voltage regulation for the load resistor R_L. The load current is only 0.283 mA, while the current through the Zener diode is 5.68 mA.

The Zener has protected the 18 kΩ load from a varying voltage source.

Lab Activity 2-2 demonstrates the behavior of a Zener diode in a live circuit.

LAB ACTIVITY **2-2**

Zener Diode and Voltage Regulation

Materials, Equipment, and Parts:

- *NTE* catalog or Internet access, www.nteinc.com

- Digital Multimeter (DMM) with test leads.

- Variable DC voltage source, 0 V to 15 V.

- Breadboard and connecting wires.

- Zener diode with a V_Z of 5 V to 12 V. Some common Zener diodes within this range are 1N4733A (5.1 V), 1N5235B (6.8 V), and 1N5240B (10 V).

- 1 kΩ fixed resistor.

- 39 kΩ fixed resistor.

Discussion Summary:

The Zener diode is a special application diode that is purposely operated in reverse bias. A Zener diode will not conduct until its reverse breakdown voltage (V_Z) is exceeded. Once the V_Z is reached, the Zener passes current but maintains a steady voltage across its terminals and also a steady voltage across anything connected in parallel with the Zener, such as a load. Thus, the Zener diode can be used for voltage regulation by protecting its load from excessive voltage.

Procedure:

SAFETY FIRST. Eye protection should always be worn when working with live voltages. Before powering on a live circuit, always check with your instructor.

1 Write the part number of the Zener diode at your workstation. Then use the *NTE* catalog or visit the website www.nteinc.com to find the NTE replacement number and the diode's V_Z.

Part number _____

NTE _____

V_Z _____

2 On the breadboard, construct the circuit below. Ensure the Zener diode is connected in reverse bias. Have your instructor check the circuit.

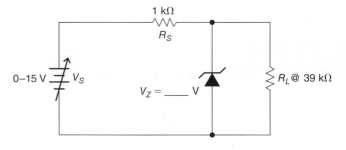

FIGURE 2-8 Zener diode circuit (© Cengage Learning 2012)

(continues)

LAB ACTIVITY 2-2

(continued)

3 Set your DMM to measure DC voltage. With the variable DC voltage source set at 3 V, turn the power on and use the voltmeter function of your DMM to measure and record the voltage across the 39 kΩ load resistor R_L. Then set the variable DC voltage source to 4 V, and again measure and record the voltage across R_L. Repeat this procedure for the following V_S values.

V_S	V_{RL}	V_S	V_{RL}
3 V	_____	10 V	_____
4 V	_____	11 V	_____
5 V	_____	12 V	_____
6 V	_____	13 V	_____
7 V	_____	14 V	_____
8 V	_____	15 V	_____
9 V	_____		

4 What was the maximum voltage recorded across the load resistor R_L?

V_{RL} (maximum) _____

5 How does the maximum voltage measured across the load resistor compare to the V_Z of the Zener diode?

6 Did the Zener diode provide voltage regulation for the 39 kΩ load resistor?

7 Build the circuit on Multisim®, Electronics Workbench®, or SPICE, and then compare the voltage measurements to the ones you just measured live. If the results are not similar, ensure you have chosen the correct values for the components.

2-3 PHOTODIODES

A **photodiode** is another special diode with many applications, including alarms, smoke detectors, and remote controls in TVs and DVD players. While an operating LED is clearly visible, photodiodes work behind the scenes.

Figure 2-9(a) shows two examples of discrete (stand-alone) two-terminal photodiodes. Photodiodes often have a clear transparent case and look like LEDs, but we'll soon discover they operate differently from LEDs.

Figure 2-9(b) shows the schematic symbol for a two-terminal photodiode. Notice that the symbol for a photodiode is similar to an LED except that the two arrows or lightning bolts point *inward*. Again, these arrows or lightning bolts are typically used in electronics to represent light. Thus, the photodiode schematic symbol has light going into the photodiode.

The operation of a photodiode is the opposite of an LED. When a forward bias voltage is applied to an LED, it gives off light. *When light strikes a photodiode, it produces a voltage.* This concept of light producing a voltage is the operating principle behind solar cells and solar panels used in satellites and home heating/cooling units.

Because LEDs give off light and photodiodes use light to produce a voltage, these two devices are used in combination for light-emitting/light-detecting systems. For example, the hand-held remote control for a TV uses an LED to emit a light signal, and a photodiode in the TV receives the light and produces a voltage signal to adjust the volume, change channels, etc. Fiber optic communication also relies on LED-to-photodiode technology. Security systems, motion detectors, and many other commercial applications use photodiodes.

A common application for the photodiode is the **optoisolator**, or optocoupler. An optoisolator uses an LED-photodiode combination that is encased in an integrated circuit. Integrated circuits will be discussed in detail in Chapter 3. For now, you just need to know that an **integrated circuit**, or "IC" or "chip," contains several electronic devices housed in one package. Also, manufacturers of ICs often put a notch or dot on the IC for identification purposes. When building a circuit with an IC, if the notch is up or at the twelve o'clock position, pin 1 is the first pin on the left. If there is no notch, then the pin closest to the dot is pin 1.

Figure 2-10(a) shows a 4N25 optoisolator IC, which is about the size of a dime. Figure 2-10(b) shows the pin configuration, internal diagram, and external connections for the IC.

Inside the optoisolator IC, you can see that there are two separate electronic devices. The left device is an LED, and the right device is actually a phototransistor (transistors are discussed in detail in Chapter 4). Since one lead is not used (pin 6), the phototransistor actually operates as a photodiode. Notice that pin 3 is labeled N.C., which means "no connection." Pin 3 serves no electrical purpose, but it provides stability when mounting the IC on a circuit board. Also, notice that there is only one pair of "light" arrows. These arrows indicate the output light for the internal LED *and* the input light for the photodiode.

Two external circuits are connected to the optoisolator IC. The circuit on the left consists of a 12 V DC voltage source, an SPST switch, and an 820 Ω current-limiting resistor R_S. The circuit on the right is made of a 5 V DC voltage source, a 270 Ω current-limiting resistor R_S, and a two-terminal LED, which in this case functions as the external load.

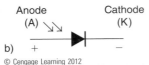

Anode (A)   Cathode (K)

a)
b)   © Cengage Learning 2012

FIGURE 2-9 Two-terminal photodiodes: a) Examples of photodiodes and b) Schematic symbol

a) Courtesy of Tracy Grace Leleux

b) © Cengage Learning 2012

FIGURE 2-10 Optoisolator: a) 4N25 integrated circuit and b) Pin configuration, internal diagram, and external connections

The optoisolator gets it supply voltage from 12 V DC applied between pin 1 and ground. To operate the optoisolator, you turn on SW1, current flows through pin 1 on the left, through the internal LED, and through pin 2 to ground. The internal LED illuminates, and the light from the LED strikes the photodiode, turning it on. When the photodiode is on, it acts like a wire to provide a path for current from the 5 V DC load voltage through pin 5 and then through pin 4 to ground. Of course, this LED-to-photodiode action occurs internally and out of sight. You can only see the result of the operation, the lighting of the *external* load LED.

The optoisolator gets its name because the external source and the external load are separated, or electrically isolated, from each other. Since the power from the 12 V source is coupled to the external LED only through the action of the internal LED and photodiode, the optoisolator is also called an optocoupler. In other words, *no light, no connection*. This application is very critical in certain digital systems like the Programmable Logic Controller (PLC), an expensive microcomputer designed to control machinery in a rugged environment. If the source

voltage gets overloaded or short-circuited, then the light beam shuts off, and the expensive load is protected.

For now, one last point about integrated circuits. As a technician, you have access only to the external pins. If the internal LED or photodiode burns out, you have to replace the entire optoisolator IC. In this regard, an IC is like the seven-segment display.

Troubleshooting Tip

The first step in troubleshooting is to do a visual inspection. If an IC chip is not working, check its power supply voltage pin and then its ground pin.

Like the Zener diode, an optoisolator can be considered a protective device. It provides electrical isolation, but it also joins or couples one external circuit to another external circuit. In Lab Activity 2-3, we'll build a circuit that contains a photodiode.

LAB ACTIVITY **2-3**

Photodiode Used in an Optoisolator Circuit

Materials, Equipment, and Parts:

- *NTE* catalog or Internet access, www.nteinc.com

- Digital Multimeter (DMM) with test leads.

- 5 V DC voltage source.

- 12 V DC voltage source.

- Breadboard and connecting wires.

- 4N25 or 4N35 optoisolator IC.

- Two-terminal LED (any color).

- SPST switch.

- 270 Ω fixed resistor.

- 820 Ω fixed resistor.

Discussion Summary:

The photodiode is a special application diode that gives off a voltage when struck by light. When a photodiode is used in combination with an LED, the device is called an optoisolator, or optocoupler. An input voltage lights the internal LED. Light from the internal LED strikes the photodiode, which provides a path for the external circuit and thus lights the *external* LED.

Procedure:

SAFETY FIRST. Eye protection should always be worn when working with live voltages. Before powering on a live circuit, always check with your instructor.

1 Write the part number of the optoisolator at your workstation. Then use the *NTE* catalog or visit the website www.nteinc.com to find the NTE replacement number, the pin configuration, and the internal diagram for the IC.

Part number _____

NTE _____

2 On the breadboard, construct the circuit below. Have your instructor check the circuit.

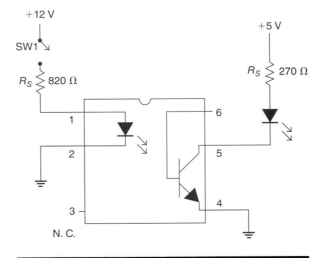

FIGURE 2-11 Optoisolator circuit (© Cengage Learning 2012)

(continues)

LAB ACTIVITY 2-3

(continued)

3 Power on the 5 V DC voltage source. The external *load* LED should not light. Power on the 12 V DC voltage source. When you turn on the SPST switch SW1, the external *load* LED should light, indicating that the optoisolator has transferred light internally from the LED to the photodiode. When you turn off the switch—simulating a problem with the source voltage—the external load LED should slowly turn off.

Does the external load LED light correctly? (circle one) Yes/No

2-4 VARACTOR DIODES

The varactor diode is another special application diode that deserves special coverage. Figure 2-12(a) shows a varactor diode. The varactor has two leads like a PN junction diode or a Zener diode. The only way to correctly identify the leads of a varactor diode is to look it up in an *NTE* catalog, on-line or on a manufacturer's spec sheet.

Figure 2-12(b) shows the schematic symbol for a two-terminal varactor. The varactor's schematic symbol looks like the symbol for a PN junction diode blended with a capacitor's schematic symbol. This is no coincidence.

Recall from Chapter 1 that every diode has a "no electron zone" between its P-type material and its N-type material. This "no electron zone" is called a depletion region because of the lack of electrons in this area. Remember, too, that forward bias decreases the barrier or depletion region in a diode, and current flows through the diode. On the other hand, reverse bias enlarges the depletion region (the "no electron zone") in a diode, and no current flows through the diode.

In AC you learned that a capacitor has a dielectric (insulator) between its leads or "plates." This insulator is also a "no electron zone." If you decrease the distance between the plates, the capacitance increases. Conversely, if you increase the distance between the plates, the capacitance decreases.

> ## *Link to Prior Learning*
> Forward bias increases capacitance, and reverse bias decreases capacitance.

Well, the varactor diode takes advantage of this phenomenon. Like the Zener diode, the varactor is *purposely hooked up in reverse bias*; the anode end of the varactor is connected to the negative side or ground of the circuit. When the reverse bias across a varactor increases, the depletion region increases, and the capacitance goes down. When the reverse bias across a varactor decreases, the depletion region decreases, and the capacitance goes up. If a varying reverse bias is applied to a varactor diode, the capacitance will also vary. Thus, a **varactor** diode acts like an automatic variable capacitor. That's why a varactor diode is also called a varicap.

Because of its varying capacitance, a varactor can be used in radio and TV circuits to provide automatic tuning. The circuitry involving varactors is complex and requires more theory that will be covered in oscillators in Chapter 9. For now, you just need to know the schematic symbol and basic operation of the varactor diode.

2-5 ADDITIONAL DIODES

Besides the LED, Zener diode, photodiode, and varactor diode discussed in this chapter, some other less-common diodes are used for particular applications. In this section, we'll briefly cover a few of these diodes that you may come across in your career as a technician.

Figure 2-13(a) shows two of these special diodes. On the left is a Schottky diode, and on

a)
Courtesy of Tracy Grace Leleux

Anode
(A)

Cathode
(K)

b)   + ───▶|◀─── –

© Cengage Learning 2012

FIGURE 2-12   Varactor diode: a) MV2109 varactor and b) Schematic symbol

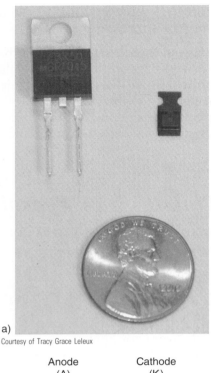

a)
Courtesy of Tracy Grace Leleux

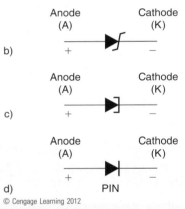

b)   Anode (A) ─►⊢ Cathode (K)   +   −

c)   Anode (A) ─►⊦ Cathode (K)   +   −

d)   Anode (A) ─►⊦ Cathode (K)   +   PIN   −

© Cengage Learning 2012

FIGURE 2-13   Special diodes: a) Schottky (left) and PIN diodes, b) Schematic symbol Schottky diode, c) Schematic symbol tunnel diode, and d) Schematic symbol PIN diode

the right is a PIN diode. The schematic symbols for the Schottky, tunnel, and PIN diodes are shown in Figures 2-13(b), 2-13(c), and 2-13(d).

The **Schottky diode**, invented by German physicist Walter Schottky, is manufactured such that it needs a very low forward bias voltage to turn it on (and thus off), as low as 0.1 V. In comparison, the germanium diode needs @ 0.3 V; the silicon diode @ 0.6 V; and the LED @ 1.5 V. Because of the low turn on/turn off voltage, the Schottky diode is used as a very high-speed electronic switch. Schottky diodes are used as rectifiers in switched-mode power supplies, as detectors in radio receiver systems, and as protection devices for systems that connect solar cells to lead-acid batteries. The Schottky diode is also known as a hot-carrier diode.

The **tunnel diode**, invented by Japanese physicist Leo Esaki, is also manufactured such that it provides high-speed switching. Like Schottky diodes, tunnel diodes are also used in communication systems, particularly high-frequency applications such as local oscillators for UHF TV tuners. The tunnel diode is also known as the Esaki diode.

The **PIN diode** is another limited-application diode. If you look at Figure 2-13(d), you'll see that the schematic symbol for a PIN diode is identical to that of the PN junction diode. To avoid confusion, the PIN diode symbol is labeled with the word *PIN*. A PIN diode acts like a capacitor when reverse biased (similar to the way a varactor behaves). When forward biased, a PIN diode functions like a RF (radio frequency) switch. PIN diodes are used as fast-acting RF switches in high-frequency test probes and in radio communication applications.

Table 2-1 lists the eight basic diodes used in electronics, their schematic symbols, and common applications. Electronic technicians should memorize the list because they will have to test, troubleshoot, and replace many of these diodes throughout their careers.

CHAPTER SUMMARY

The light-emitting diode (LED) is one of several special-purpose diodes. When voltage is applied to an LED, it gives off light, making the LED useful for panel indicators, automotive headlights, and displays on calculators, billboards, computers, and clocks. Today's LEDs provide more illumination, consume less power, and are more durable than traditional incandescent (metal filament) light bulbs.

The Zener diode provides protection for a load through voltage regulation. The Zener is connected in reverse bias, and unlike the PN junction diode and LED, it does not start conducting until the reverse bias voltage exceeds a certain value called the V_Z. The Zener then passes current but maintains a steady voltage across its terminals and any load connected in parallel with it, such as a washing machine motor or computer CPU.

The photodiode is a special application diode that turns on and produces a voltage when struck

TABLE 2-1 Eight Basic Diodes
(© Cengage Learning 2012)

Label	Schematic	Application	
PN Junction		Rectifier	
Zener		Voltage regulator	√
LED		Panel indicator	
Varactor (variable capacitor)		Tuned circuits	√
PIN	PIN	High-Frequency Test Probes	
Schottky		Switched-Mode Power Supplies and Electric Cars	
Tunnel		Local Oscillators for UHF TV Tuners	
Photodiode		Remote Controls for TVs and DVD Players; Security Systems	

(√) *Diodes used in Reverse Bias*

by light. When a photodiode is used in combination with an LED, the device is called an optoisolator, or optocoupler. Common applications for photodiodes and optoisolators include light-emitting/light-detecting systems such as hand-held remote controls for TVs, DVD players, fiber optic, and security systems.

Like the Zener diode, varactor diodes are connected in reverse bias. They act like variable capacitors, so they are used in radio and TV circuits to provide automatic tuning.

Other less-common special-purpose diodes include the Schottky diode, which acts like a very high-speed electronic switch for power supplies and also as a detector in radio receivers; the tunnel diode, which provides high-speed switching for UHF TV tuners; and the PIN diode, which can serve the dual role of capacitor and RF (radio frequency) switch.

CHAPTER EQUATIONS

Current-limiting resistor

(Equation 2-1) $\quad R_S = \dfrac{V_S - V_F}{I_F}$

V_S = source voltage

V_F = voltage drop across LED in forward bias

I_F = current through LED in forward bias

Zener off

(Equation 2-2) $\quad I_T = \dfrac{V_T}{R_T}$

$R_T = R_S + R_L$

Zener on

(Equation 2-3) $\quad I_T = \dfrac{V_{RS}}{R_S}$

V_{RS} = voltage across current-limiting resistor R_S

Current through a Zener diode

(Equation 2-4) $\quad I_Z = I_T - I_{RL}$

I_Z = current through Zener diode

I_{RL} = current through load resistor

CHAPTER REVIEW QUESTIONS

Chapter 2-1

1. List two reasons why LEDs have replaced miniature incandescent (metal filament) light bulbs and lamps in many applications.

2. How do you identify the cathode of an LED?

3. Draw the schematic symbol for a two-terminal LED and explain its operation.

4. Explain the procedure for testing a two-terminal LED to determine if it is good.

5. What is the typical voltage drop across an LED in the forward bias condition?

6. Explain the difference between common-anode and common-cathode seven-segment LED displays.

7. Why must every LED have a resistor in series with it?

8. If pins are missing on a seven-segment LED, does it mean the display is necessarily bad?

9. What segments of a seven-segment LED display are needed to illuminate the number *5*?

10. If the *b* segment of a seven-segment LED does not light, what are the two possible causes?

Chapter 2-2

11. What is the primary application for a Zener diode?

12. Draw and label the schematic symbol for a Zener diode.

13. Explain the basic operation of a Zener diode.

14. Use an *NTE* catalog or the website www.nteinc.com to find the V_Z for diode part number 1N4744A.

15. Explain the procedure for testing a Zener diode to determine if it is good.

16. If a Zener diode with a V_Z of 12 V is on, what is the voltage across a 27 kΩ resistor connected in parallel with the Zener?

17. Determine V_{RL}, V_{RS}, I_T, I_{RL}, and I_Z for the circuit of Figure 2-14.

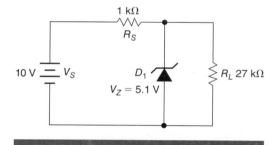

FIGURE 2-14 Zener diode in a DC circuit (© Cengage Learning 2012)

Chapter 2-3

18. Draw and label the schematic symbol for a two-terminal photodiode.

19. Explain the basic operation of a photodiode.

20. What are the two major electronic devices in an optoisolator or optocoupler?

Chapter 2-4

21. Which two diodes discussed in this chapter are meant to be connected in reverse bias to operate correctly?

22. Draw and label the schematic symbol for a varactor diode and explain its operation.

23. A varactor diode behaves like which AC component?

24. Name one common application for a varactor diode.

Chapter 2-5

25. What diode discussed in this chapter has a schematic symbol identical to the PN junction diode?

26. Name two diodes discussed in this chapter that can perform as high-speed electronic switches.

27. Draw and label the schematic symbol for the tunnel diode.

Rectifiers and Power Supplies

OBJECTIVES   *Upon completion of this chapter, you should be able to:*

- List and explain the stages of a power supply.
- Explain the operation of a half-wave rectifier.
- Distinguish conventional current flow from electron flow.
- Construct a half-wave rectifier and measure the average (pulsating DC) load voltage.
- Use an electrolytic capacitor to provide filtering.
- Explain the operation of a full-wave center-tapped rectifier and calculate the average (pulsating DC) load voltage.
- Construct a full-wave center-tapped rectifier to calculate and measure the average (pulsating DC) load voltage.
- Explain the operation of a full-wave bridge rectifier and calculate the average (pulsating DC) load voltage.

- Construct a full-wave bridge rectifier to calculate and measure the average (pulsating DC) load voltage.
- List the advantages and disadvantages of integrated circuits (ICs).
- Distinguish between through-hole technology and surface mount technology used to manufacture ICs.
- Identify the pins of an IC.
- Construct and analyze a complete power supply.
- Calculate the voltage regulation percentage for a power supply.
- Explain the operation of a voltage doubler.
- Construct a voltage doubler to calculate and measure the output voltage.

MATERIALS, EQUIPMENT, AND PARTS

Materials, equipment, and parts needed for the lab experiments in this chapter are listed below:

- *NTE* catalog or Internet access, www.nteinc.com
- Digital Multimeter (DMM) with test leads.
- Dual trace oscilloscope w/BNC-to-alligator leads.
- 6.3 V AC power supply.
- 12.6 V AC power supply with center-tap.

- Breadboard and connecting wires.
- 1N4001 diode or equivalent (4).
- 1N4742A Zener diode (12 V_Z) *or* a Zener close to 12 V value.
- 1 µF electrolytic capacitor, 47 µF electrolytic capacitor (2), and 100 µF electrolytic capacitor.

- 270 Ω fixed resistor, 330 Ω fixed resistor, 10 kΩ fixed resistor, 12 kΩ fixed resistor, and 82 kΩ fixed resistor.
- [Optional] IC bridge rectifier such as DB102 (NTE 5332), DF02 (NTE 5332), *or* one that can handle 1 A for safety sake.

GLOSSARY OF TERMS

Power supply The unit in most electronic systems that provides the voltages to drive the subsystems

Transformer An electrical device typically used in a power supply to reduce or "step-down" the incoming line voltage

Filtering A process in a power supply where a capacitor or capacitors remove the pulsating or varying portion of a rectified waveform

Regulation A process in a power supply in which an electronic device provides a steady DC voltage to the load

Conventional flow The view that current flows from the positive side of a voltage source to the negative side

Electron flow The view that current flows from the negative side of a voltage source to the positive side

Half-wave rectifier A semiconductor circuit that changes AC to pulsating DC by reproducing across a load either the positive or negative alternation of an AC input signal

Full-wave center-tapped rectifier A semiconductor circuit made of two diodes and a center-tapped transformer that changes AC to pulsating DC by reproducing across a load both alternations of an AC input signal

Full-wave rectifier A semiconductor circuit that changes AC to pulsating DC by reproducing across a load both alternations of an AC input signal

Full-wave bridge rectifier A semiconductor circuit or device made of four diodes that changes AC to pulsating DC by reproducing across a load both alternations of an AC input signal

Discrete device An individual electronic device, such as a resistor, capacitor, or diode

Through-hole technology The manufacturing process of printed circuit boards where the pins or leads of an electronic device are inserted directly into pre-drilled holes in a circuit board and then soldered on the opposite side to circular pads

Surface Mount Technology (SMT) The manufacturing process of printed circuit boards where the pins or leads of an electronic device are soldered directly on the surface, or one side, of the board

Voltage regulation The change in output voltage of a power supply from a no-load to full-load condition

Voltage multiplier An electronic circuit that uses a combination of diodes and capacitors to change AC to pulsating DC *and* increase the output voltage of an AC voltage source

Voltage doubler An electronic circuit that uses two diodes and three capacitors to double the peak output voltage of an AC voltage source while changing it to pulsating DC

Voltage tripler An electronic circuit that uses three diodes and three capacitors to triple the peak output voltage of an AC voltage source while changing it to pulsating DC

3-1 POWER SUPPLY OVERVIEW

Power supplies are found in almost every electronic system, including clock radios, TVs, stereos, and laptop computers. The **power supply** in a system provides the voltages that drive the subsystems. Since every device is directly or indirectly connected to the power supply, it's important for a technician to identify and troubleshoot the electronic devices within a power supply. In fact, before replacing the motherboard of a computer, you should check the power supply voltages to see if they are correct.

Diodes play an important role in power supplies along with other electronic devices. Before we discuss in detail the diode's role in a power supply, we'll look at the various stages of a typical power supply.

Figure 3-1 shows a simplified block diagram of the stages of a power supply. The first stage of any power supply is the Voltage Reduction stage. In this stage, the incoming AC line voltage is reduced to a smaller amount to be used by the other stages. A **transformer** is used to reduce or "step-down" the voltage. In the power supply of Figure 3-1, the transformer reduces the incoming 120 V AC to 10 V AC. Remember, a transformer doesn't change the voltage from AC to DC: it just changes the amplitude, or amount. Transformers were discussed in detail in your AC class. We'll review them during this chapter.

The second stage is the Rectification stage, which is made of one or more PN junction diodes. During this stage, the incoming AC voltage is "rectified" or changed to pulsating DC. It is now DC because one of its alternations has been cut off

or clipped. It is called "pulsating" because the voltage still varies in amplitude, or amount. In the case of Figure 3-1, the negative alternation has been clipped. The Rectification stage will be discussed in detail in this chapter.

The third stage is called the Filtering stage. **Filtering** means to remove the pulsating or varying portion of the waveform. In this stage, an electrolytic capacitor charges rapidly and discharges slowly so that the pulsating DC voltage is flattened to resemble a DC voltage, which looks like a straight line on an oscilloscope.

The final stage is the **Regulation** stage. The Zener diode that we discussed in detail in Chapter 2 is used here to provide a steady DC voltage to the load. The load is what consumes the power in any electronic circuit such as a computer's motherboard, a vacuum cleaner motor, or a television screen.

The input to a power supply is AC, and the output is filtered and regulated DC. This process occurs in a laptop "adapter," a stereo system, a television—practically anything you plug into your AC wall outlet. Many electronic devices, such as the CPU of a computer, use DC voltages with very strict tolerances, so the power supply plays an important role in providing the correct voltages to a load. In this chapter, we'll focus first on the Rectification stage.

3-2 HALF-WAVE RECTIFIERS

We mentioned in Chapter 1 that the PN junction diode is most commonly used as a rectifier; that is, it changes alternating current (AC) to pulsating

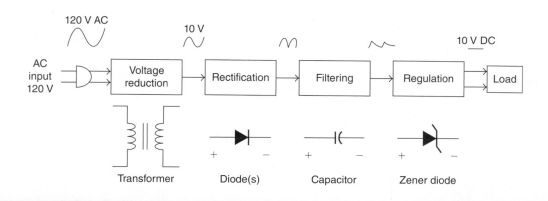

FIGURE 3-1 Block diagram of power supply stages (© Cengage Learning 2012)

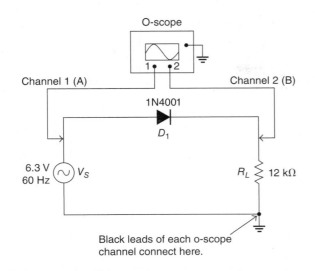

FIGURE 3-2 Half-wave rectifier circuit (© Cengage Learning 2012)

direct current (DC). Rectifier circuits are used in a variety of power supplies.

Figure 3-2 shows a half-wave rectifier circuit with a 6.3 V AC source @ 60 Hz, the standard frequency coming from an AC outlet in North America. AC power supplies—like those in most electronic educational kits or labs—commonly provide either 6.3 V AC or 12.6 V AC or both. The circuit also contains a 1N4001 or equivalent silicon diode and a 12 kΩ fixed resistor.

The 12 kΩ resistor is called the load resistor and is labeled R_L. The load resistor R_L in this case is used to show the output of the Rectification *stage*; it is not meant to represent the *load* at the output of the entire power supply. Because the 12 kΩ resistor is in series with the diode, it also acts as a current-limiting resistor.

The 6.3 V AC source is on the left; the anode of the diode connects to the top side of the source; the cathode of the diode connects to the top, or the hot side, of the 12 kΩ load resistor; and the bottom of the 12 kΩ fixed resistor connects to the bottom, or common side, of the 6.3 V AC source. A dual trace oscilloscope is connected to the circuit. The red lead of Channel 1, or A, of the oscilloscope is connected to the 6.3 V AC source and will show the full AC *input* sine wave on the oscilloscope screen. The red lead of Channel 2, or B, of the oscilloscope is connected across the 12 kΩ load resistor R_L and will show the *output* waveform appearing across R_L. The black leads of each o-scope channel connect to the "ground" point at the bottom of the

circuit. Both channels of the o-scope are set to measure AC.

Before we examine the circuit closely, let's talk briefly about the two approaches to the direction of current flow: conventional current flow and electron flow. **Conventional flow** claims (falsely) that current flows from the positive side of a voltage source to its negative side. This approach was established in the 1800s by electronic pioneers such as Georg Simon Ohm and Gustav Kirchhoff, who didn't know of the existence of the electron. As a result, schematic symbols like the one used for a diode show current flowing in the direction of the arrow; that is, from anode to cathode or from positive to negative.

In the early 1900s, physicists proved that the flow of current was really the movement of electrons from atom to atom. This means that current flows from the negative side of a voltage source to the positive side of the voltage source or from cathode to anode in a diode. This is called **electron flow**. It really doesn't matter which approach you embrace because current flow is instantaneous anyway—when you turn on a switch, the current's done flowing before you can snap your fingers. To keep things simple, we'll use *conventional current flow* in the analysis of circuits; that is, in the direction of the diode arrow.

Now, back to the circuit. Remember from your AC course that current from an AC voltage source travels in one direction, and then the current reverses direction. For learning purposes only, we are going to "freeze" the AC voltage source. When the 6.3 V AC source is positive on the top and negative on the bottom as shown in Figure 3-3(a), the diode is forward biased—positive to anode and negative to cathode—and current flows through the diode and through the 12 kΩ load resistor R_L. The positive alternation of the 6.3 V AC source appears across the load resistor R_L and on Channel 2 of the oscilloscope screen as shown in Figure 3-3(b).

When the 6.3 V AC source is negative on the top and positive on the bottom as shown in Figure 3-4(a), the diode is reverse biased—negative to anode and positive to cathode—and no current flows through the diode or the load resistor R_L. The waveform taken across the 12 kΩ load resistor as shown in Figure 3-4(b) is indicated by a blue dotted line where the negative alternation should be. The diode has provided rectification; it has changed the AC signal to a pulsating DC signal. We consider the output signal pulsating DC because it still varies

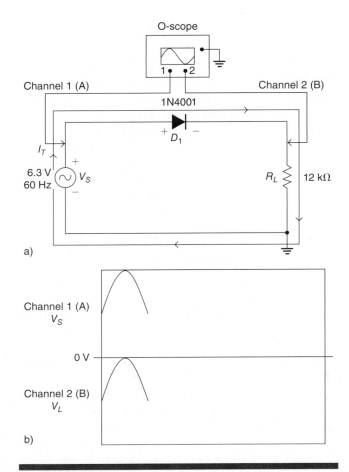

FIGURE 3-3 Half-wave rectifier with diode conducting: a) Schematic and b) O-scope display (© Cengage Learning 2012)

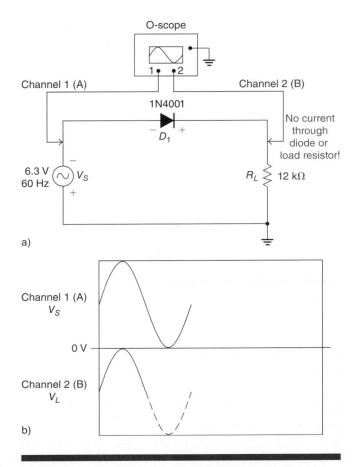

FIGURE 3-4 Half-wave rectifier with diode not conducting: a) Schematic and b) O-scope display (© Cengage Learning 2012)

in amplitude like an AC waveform, but it has only one alternation like a DC waveform.

Of course, in real life we can't freeze an AC waveform. Figure 3-5 shows the waveforms as you would see them on an o-scope. The top of Figure 3-5 shows the AC waveform going into the circuit, and the bottom shows the waveform that appears across the 12 kΩ load resistor. Notice that the negative alternation of the AC sine wave is missing. This is because the diode passes current only on the positive alternation; that is, when the diode is forward biased.

This circuit is called a positive **half-wave rectifier** because only the positive alternation of the AC input signal, or wave, is reproduced across the load resistor. The circuit is also called a negative clipper because it eliminates or clips the negative alternation of the AC input signal.

To create a negative half-wave rectifier, just flip the diode. In the negative half-wave rectifier, the

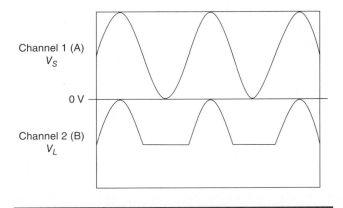

FIGURE 3-5 Real-life oscilloscope display of half-wave rectifier (© Cengage Learning 2012)

negative alternation of the AC input signal is reproduced across the load resistor. The negative half-wave rectifier is also called a positive clipper because it eliminates or clips the positive alternation of the AC input signal.

We can demonstrate the Filtering stage of a power supply by using an electrolytic capacitor. Again, during this stage, the pulsating or varying DC (often called *ripple*) is removed or "filtered." In Figure 3-6(a), a 1 µF electrolytic capacitor has been inserted in parallel with the load. The electrolytic capacitor charges rapidly and discharges slowly so that the pulsating DC voltage is flattened to resemble a DC voltage. This effect is shown in the bottom waveform of Figure 3-6(b). The greater the value of the electrolytic capacitor, the more flattened the waveform. Remember, a DC voltage looks like a straight line on an oscilloscope because its voltage does not vary like an AC voltage.

When using an electrolytic capacitor in a live circuit, it's important that you connect it correctly. Remember, an electrolytic capacitor has a positive side and a negative side. The negative side is clearly marked on the capacitor with a colored band and usually a negative sign (−). The negative side *must* be connected to the bottom of the load resistor. Paying attention to the positive and negative leads of an electronic device when connecting it to a circuit is called "observing polarity." Hooking up an electrolytic capacitor incorrectly can cause it to explode!

Half-wave rectifiers are rarely used in commercial applications because they waste power in the form of a lost alternation. However, they are a great learning tool because they are easy to build and demonstrate rectification. In Lab Activity 3-1, we will build and see the effects of half-wave rectification.

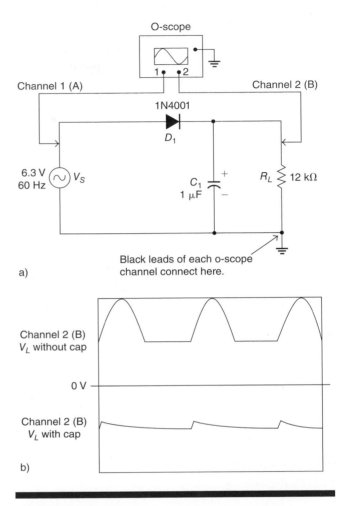

FIGURE 3-6 Half-wave rectifier circuit filter: a) Schematic and b) O-scope display across R_L without filter capacitor and with filter capacitor (© Cengage Learning 2012)

Positive Half-Wave Rectifier

Materials, Equipment, and Parts:

- Digital Multimeter (DMM) with test leads.

- Dual trace oscilloscope w/BNC-to-alligator leads.

- 6.3 V AC power supply.

- Breadboard and connecting wires.

- 1N4001 diode or equivalent.

- 1 µF electrolytic capacitor.

- 100 µF electrolytic capacitor.

- 12 kΩ fixed resistor.

Discussion Summary:

A half-wave rectifier uses a diode to convert AC to pulsating DC. On the positive alternation of the AC source, the diode is forward biased, current flows through the diode and load resistor, and the positive alternation appears across the load resistor R_L. On the negative alternation of the AC source, the diode is reverse biased, no current flows through the diode or load resistor, and thus no waveform appears across the load resistor.

Procedure:

SAFETY FIRST. Eye protection should always be worn when working with live voltages. Before powering on a live circuit, always check with your instructor.

1 Build the circuit shown in Figure 3-7. Connect the positive lead of the oscilloscope's Channel 1 to the hot (more positive) side of the AC power supply. Connect the positive lead of the o-scope's Channel 2 to the top of the 12 kΩ load resistor. Connect both o-scope negative leads to a wire inserted at the bottom of the load resistor.

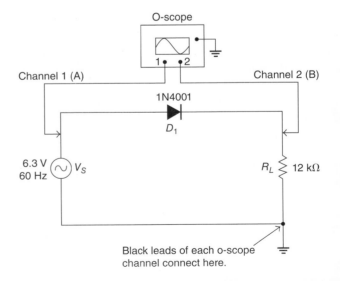

FIGURE 3-7 Half-wave rectifier circuit (© Cengage Learning 2012)

(continues)

LAB ACTIVITY 3-1

(continued)

2 Set your DMM to measure AC voltage. Use your DMM to measure and record the AC power supply voltage (@ 6.3 V).

$V_S = \underline{\hspace{2cm}}$

3 Power on the circuit. Set your Mode switch to dual. Adjust your oscilloscope volts/division settings and time/division settings to get the input and output waveforms to appear like those of Figure 3-5.

Link to Prior Learning

An oscilloscope (o-scope) measures voltage over a period of time. Once you connect the o-scope input leads to a voltage source, the voltage is at the o-scope display. It's your job to make the waveform appear clearly by adjusting the various settings, particularly volts/division and time/division settings.

4 Set your DMM to measure DC voltage. Remember, the output voltage across the load resistor is now *pulsating DC*, so you have to set your meter to measure DC voltage. Measure and record the voltage across the 12 kΩ load resistor.

$V_{RL} = \underline{\hspace{2cm}}$

5 Compare the input voltage from Step 2 to the output voltage from Step 4. Why is there such a difference?

6 Power off the circuit. Flip the diode so that the cathode is facing to the left toward the hot (more positive) side of the AC power supply. Now you've created a negative half-wave rectifier circuit. Power on and observe the o-scope display. Has anything changed?

7 Power off the circuit. Flip the diode to its original position—anode to the left toward the hot (more positive) side of the AC power supply.

8 *Observing polarity*, carefully insert a 1 µF electrolytic capacitor in parallel with and to the left of the 12 kΩ load resistor as shown in Figure 3-6.

9 Have the instructor check your circuit.

LAB ACTIVITY 3 - 1

10 Power on. What happened to the waveform across the 12 kΩ R_L? Did the capacitor provide filtering? Power off.

11 *Observing polarity*, carefully replace the 1 µF electrolytic capacitor with a 100 µF electrolytic capacitor, ensuring it is in parallel with and to the left of the 12 kΩ load resistor.

12 Have the instructor check your circuit.

13 Power on. What happened to the waveform across the 12 kΩ R_L? Did the capacitor provide filtering? Is the waveform produced by the 100 µF capacitor different from that produced by the 1 µF capacitor in Step 10? Power off.

14 Build the circuit on Multisim®, Electronics Workbench®, or SPICE and then compare the results.

3-3 FULL-WAVE CENTER-TAPPED RECTIFIER

The half-wave rectifiers we built in Lab Activity 3-1 are rarely used in commercial applications because they waste power: half of the power supply voltage doesn't make it to the load. The more practical approach would be to use a circuit that uses the full power supply voltage and thus wastes less power.

The **full-wave center-tapped rectifier** shown in Figure 3-8 is more efficient than the half-wave rectifier. The full-wave center-tapped transformer has two diodes connected to a center-tapped transformer and a 12 kΩ load resistor R_L. Again, the 12.6 V AC feeding the two diodes is typically provided by electronic educational kits or lab power supplies. The 12 kΩ resistor is in series with both diodes and acts as a current-limiting resistor.

A dual trace oscilloscope is connected to the circuit. The red lead of Channel 1, or A, of the oscilloscope is connected to the 12.6 V AC transformer secondary and will show the full AC *input* sine wave applied to the rectifier circuit. The red lead of Channel 2, or B, of the oscilloscope is connected across the 12 kΩ load resistor R_L and will show the *output* waveform appearing across R_L. Both channels of the o-scope are set to measure AC. The black leads of each o-scope channel, the center-tap or 0 V line from the transformer secondary, and the bottom of the load resistor all connect to "ground." I tell my students that ground is just a common point. When building the circuit live, pick a hole on the breadboard, connect a black wire to it, and then connect all points to be grounded to the black wire.

Again, for learning purposes only, we are going to "freeze" the AC voltage source. When the 12.6 V AC secondary is *positive* on the top with respect to the grounded center-tap as shown in Figure 3-9(a), D_1 is forward biased—positive to anode and negative to cathode—and current flows through the diode, through the 12 kΩ load resistor R_L, and through ground back to the center-tap. The positive alternation of the 12.6 V AC source appears across the load resistor R_L and on Channel 2 of the oscilloscope screen as shown in Figure 3-9(b). Notice that current doesn't go through D_2 because it is reverse biased, and thus no current flows through it.

Now, when the 12.6 V AC secondary is positive on the *bottom* with respect to the grounded center-tap as shown in Figure 3-10(a), D_2 is forward biased—positive to anode and negative to cathode—and current flows through the diode, through the 12 kΩ load resistor R_L, and through ground back to the center-tap. The negative alternation of the 12.6 V AC source appears across the load resistor R_L and on Channel 2 of the

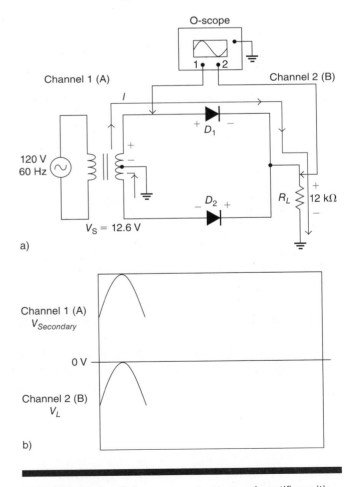

a)

b)

FIGURE 3-9 Full-wave center-tapped rectifier with D_1 conducting: a) Schematic and b) O-scope display
(© Cengage Learning 2012)

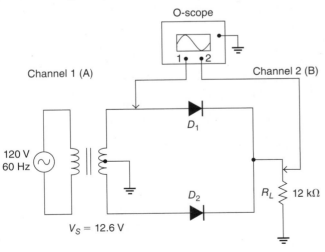

FIGURE 3-8 Full-wave center-tapped rectifier
(© Cengage Learning 2012)

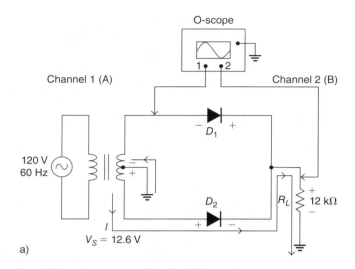

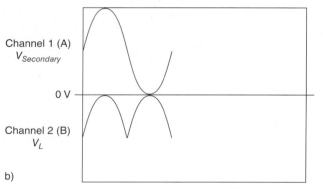

FIGURE 3-10 Full-wave center-tapped rectifier with D_2 conducting: a) Schematic and b) O-scope display (© Cengage Learning 2012)

oscilloscope screen as shown in Figure 3-10(b). Current doesn't go through D_1 because it is now reverse biased.

The diode has provided full-wave rectification; it has changed the AC signal to a pulsating DC signal. We consider the output signal pulsating DC because it still varies in amplitude like an AC waveform but has no negative alternation. The negative alternation from the 12.6 V AC input to the circuit has been changed to a positive waveform at the 12 kΩ load resistor because for both alternations the current goes through the load resistor in the same *direction*. Remember, one of the defining characteristics of AC is that it reverses direction.

Of course, in real life we can't freeze an AC waveform. Figure 3-11 shows the waveforms as you would see them on an o-scope. The top of Figure 3-11 shows the AC waveform going into the circuit, and the bottom shows the waveform that appears across the 12 kΩ load resistor.

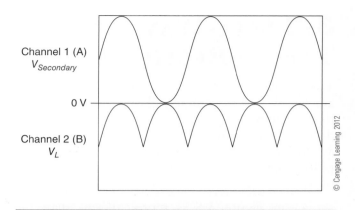

FIGURE 3-11 Real-life oscilloscope display of full-wave center-tapped rectifier (© Cengage Learning 2012)

This circuit is called a **full-wave rectifier** because both the positive alternation and negative alternation of the AC input signal are reproduced across the load resistor. Full-wave center-tapped rectifiers are used in many power supply applications. They are an improvement over the half-wave rectifier because they provide more load voltage than the half-wave rectifier. How much load voltage? you ask. Let's figure it out.

Since rectifiers change AC to pulsating DC, we need to review a few concepts we learned way back in AC before we can determine the pulsating DC voltage across the load.

Link to Prior Learning

The peak voltage in an AC circuit equals the RMS voltage divided by .707

$$V_P = \frac{V_{RMS}}{.707}$$

In a full-wave rectifier, the average (pulsating DC) load voltage equals .637 multiplied by the V_P across the load resistor.

$$V_{AVG} = .637 \ (V_P)$$

So, for the full-wave rectifier of Figure 3-8, we can determine the V_P across the transformer secondary.

$$V_P \text{ (across secondary)} = \frac{V_{RMS}}{.707} = \frac{12.6 \text{ V}}{.707} = 17.82 \text{ V}$$

Now, we need a new equation to determine the peak voltage across the 12 kΩ load resistor. Looking back at Figures 3-9 and 3-10, we see that for each alternation, only one-half of the center-tapped transformer is being used. Also, for each alternation, only one diode is conducting and thus drops @ 0.6 V because each diode is silicon. (Remember, assume a diode is silicon if it's not specifically labeled germanium.) This leads to an equation to help determine the actual peak voltage that appears across the 12 kΩ load resistor.

EQUATION 3-1

Full-wave center-tapped rectifier (without filter capacitor)

$$V_P \text{ (across the load)} = \frac{V_P \text{ (secondary)}}{2} - .6 \text{ V}$$

$$V_P = \text{peak voltage}$$

So, for Figure 3-8, V_P (across the load) = $\frac{17.82 \text{ V}}{2} - .6 \text{ V} = 8.31 \text{ V}$

Once we have the peak voltage across the load, we can calculate the average value; that is, the pulsating DC voltage.

EQUATION 3-2

Full-wave rectifier (without filter capacitor)

$$V_{AVG} = .637 \ (V_P)$$

$$V_{AVG} = \text{pulsating DC voltage across the load}$$

$$V_P = \text{peak voltage across the load}$$

So, for Figure 3-8, $V_{AVG} = .637 \ (V_P) = .637 \ (8.31 \text{ V}) = 5.29 \text{ V}$

This is the voltage you should measure if you use your DMM set to the DC voltage setting. It may be a little lower or higher since we estimated the diode voltage drops @ 0.6 V. Remember, once an AC signal is applied to a rectifier circuit, it is no longer AC—it is pulsating *DC*.

Let's do another example to reinforce the calculations we've just learned.

For Figure 3-12, you should measure @ 6.82 V across the load resistor R_L if you use your DMM set to the DC voltage setting. It may be a little lower or higher since we estimated the diode voltage drops @ 0.6 V.

Let's build a live version of the full-wave center-tapped rectifier.

EXAMPLE 1

Situation

Determine V_{AVG} for the circuit of Figure 3-12.

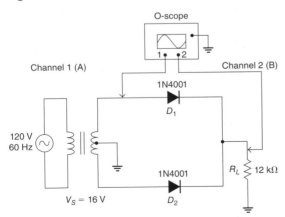

Solution

For Figure 3-12,

$$V_P \text{ (across secondary)} = \frac{V_{RMS}}{.707} = \frac{16 \text{ V}}{.707}$$

$$= 22.63 \text{ V}$$

$$V_P \text{ (across the load)} = \frac{22.63 \text{ V}}{2} - .6 \text{ V}$$

$$= 10.71 \text{ V}$$

$$V_{AVG} = .637 \ (V_P) = .637 \ (10.71 \text{ V}) = 6.82 \text{ V}$$

FIGURE 3-12 Full-wave center-tapped rectifier circuit (© Cengage Learning 2012)

LAB ACTIVITY 3-2

Full-Wave Center-Tapped Rectifier

Materials, Equipment, and Parts:

- Digital Multimeter (DMM) with test leads.

- Dual trace oscilloscope w/BNC-to-alligator leads.

- 12.6 V AC power supply with center-tap.

- Breadboard and connecting wires.

- 1N4001 diode or equivalent (2).

- 1 µF electrolytic capacitor.

- 100 µF electrolytic capacitor.

- 12 kΩ fixed resistor.

Discussion Summary:

A full-wave center-tapped rectifier uses two diodes to convert AC to pulsating DC. On the positive alternation of the AC source, one diode is forward biased, current flows through the diode and load resistor, and the positive alternation appears across the load resistor R_L. On the negative alternation of the AC source, the other diode is forward biased, current flows through the diode and load resistor, and the negative alternation appears as a positive voltage across the load resistor R_L.

Procedure:

SAFETY FIRST. Eye protection should always be worn when working with live voltages. Before powering on a live circuit, always check with your instructor.

1 Build the circuit shown in Figure 3-13. Connect the red lead of Channel 1, or A, of the oscilloscope to one side of the 12.6 V AC transformer secondary. Connect the red lead of Channel 2, or B, of the o-scope to the top of the 12 kΩ load resistor. Connect the black leads of each o-scope channel, a wire from the center-tap or 0 V line from the transformer secondary, and a wire from the bottom of the load resistor to one point (hole) on the breadboard.

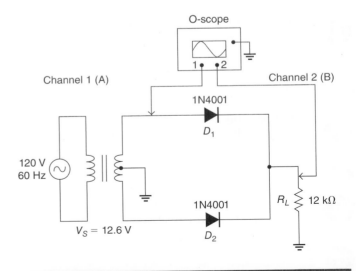

FIGURE 3-13 Full-wave center-tapped rectifier circuit (© Cengage Learning 2012)

(continues)

LAB ACTIVITY 3-2

(continued)

2   Set your DMM to measure AC voltage. Use your DMM to measure and record the AC power supply voltage (@ 12.6 V).

$V_S =$ _____

3   Use the following AC equation to determine the peak voltage across the transformer secondary:

V_P (across secondary) $= \dfrac{V_{RMS}}{.707} =$

V_P (across secondary) $=$ _____

4   Use Equation 3-1 to determine the peak voltage across the load resistor.

(***Equation 3-1***) V_P (across the load) $= \dfrac{V_P \text{ (secondary)}}{2} - .6 \text{ V} =$

V_P (across the load) $=$ _____

5   Use Equation 3-2 to calculate the average or pulsating DC voltage across the load.

(***Equation 3-2***) $V_{AVG} = .637 \ (V_P) =$

$V_{AVG} =$ _____

6   Have your instructor check your circuit.

7   Power on the circuit. Set your Mode switch to dual. Adjust your oscilloscope volts/division settings and time/division settings to get the input and output waveforms to appear like those of Figure 3-11.

8   Set your DMM to measure DC voltage. Remember, the output across the load resistor is now *pulsating DC*, so you have to set your meter to measure DC voltage. Measure and record the average (pulsating DC) voltage across the 12 kΩ load resistor.

$V_{AVG} =$ _____

9   Compare the calculated average voltage across the load resistor (Step 5) to the measured average voltage across the load resistor (Step 8). Explain any differences.

10   Power off the circuit.

11   *Observing polarity*, carefully insert a 1 µF electrolytic capacitor in parallel with and to the left of the 12 kΩ load resistor.

12   Have the instructor check your circuit.

LAB ACTIVITY 3-2

13 Power on. What happened to the waveform across the 12 kΩ R_L? Did the capacitor provide filtering?

14 Power off the circuit.

15 *Observing polarity*, carefully replace the 1 μF capacitor with the 100 μF capacitor.

16 Have the instructor check your circuit.

17 Power on. What happened to the waveform across the 12 kΩ R_L? Did the capacitor provide filtering? How does the voltage waveform using the 100 μF capacitor compare to the waveform using the 1 μF capacitor from Step 11?

18 Build the circuit on Multisim®, Electronics Workbench®, or SPICE and then compare the results.

3-4 FULL-WAVE BRIDGE RECTIFIERS

The full-wave center-tapped rectifier is a noticeable improvement over the half-wave rectifier because it uses both alternations of the AC input voltage and thus wastes less power. However, a center-tapped transformer adds cost to the circuit. The answer to this problem is the **full-wave bridge rectifier**.

Figure 3-14 shows a full-wave bridge rectifier. Notice that the transformer doesn't use a center-tap and that the circuit uses four diodes connected in a diamond shape. One lead of the 12.6 V AC secondary connects to the top of the diamond, and the other lead of the 12.6 V AC secondary connects to the bottom of the diamond. The right end of the diamond is connected to the top of the 12 kΩ load resistor. The bottom of the load resistor and the left end of the diamond are connected to each other (a common point on the breadboard).

A dual trace oscilloscope is connected to the circuit. The red lead of Channel 1, or A, of the oscilloscope is connected to the top of the 12.6 V AC transformer secondary, and the black lead of Channel 1, or A, is connected to the bottom of the 12.6 V AC secondary; Channel 1 will show the full AC *input* sine wave applied to the rectifier circuit. The red lead of Channel 2, or B, of the oscilloscope is connected across the 12 kΩ load resistor R$_L$ and will show the *output* waveform appearing across R$_L$. Both channels of the o-scope are set to measure AC. The black lead of

Channel 2, or B, is connected to the bottom of the load resistor.

Again, for learning purposes only, we are going to "freeze" the AC voltage source. When the 12.6 V AC secondary is *positive* on the top with respect to bottom as shown Figure 3-15(a), D$_1$ is forward biased—positive to anode and negative to cathode—and current flows through D$_1$, down through the 12 kΩ load resistor R$_L$, back up into the left of the diamond, down through D$_3$ (also forward biased), and back up to the more negative side of the secondary. The positive alternation of the 12.6 V AC secondary appears across the load resistor R$_L$ and on Channel 2 of the oscilloscope screen as shown in Figure 3-15(b). So, for the positive alternation of the transformer secondary, two diodes (D$_1$ and D$_3$) are forward biased and thus pass current. Notice that current doesn't go through D$_2$ or D$_4$ because they are reverse biased and thus no current flows through them.

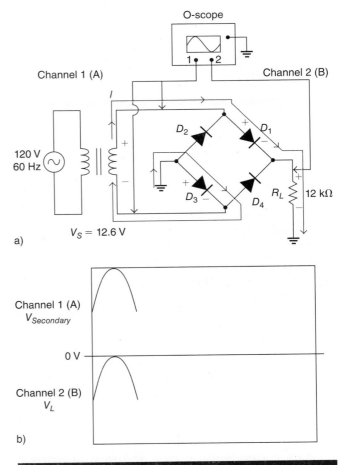

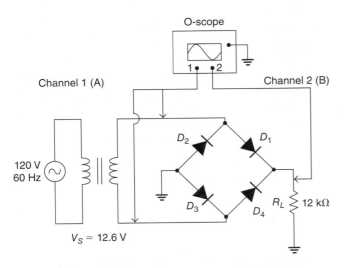

FIGURE 3-14 Full-wave bridge rectifier circuit (© Cengage Learning 2012)

FIGURE 3-15 Full-wave bridge rectifier circuit with D$_1$ and D$_3$ conducting: a) Schematic and b) O-scope display (© Cengage Learning 2012)

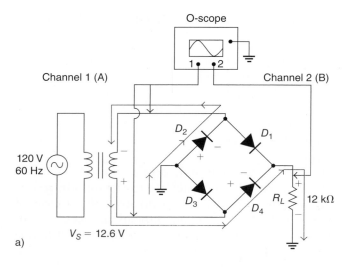

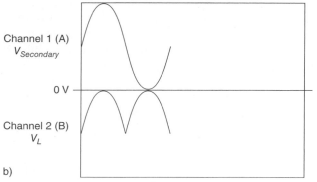

FIGURE 3-16 Full-wave bridge rectifier circuit with D_2 and D_4 conducting: a) Schematic and b) O-scope display (© Cengage Learning 2012)

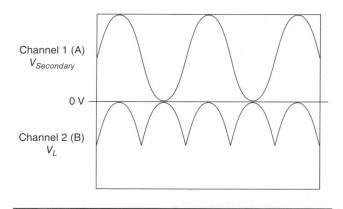

FIGURE 3-17 Real-life oscilloscope display of full-wave bridge rectifier (© Cengage Learning 2012)

Now, when the 12.6 V AC secondary is positive on the *bottom* with respect to the top as shown in Figure 3-16(a), D_4 is forward biased—positive to anode and negative to cathode—and current flows through D_4, down through the 12 kΩ load resistor R_L, back up into the left of the diamond, up through D_2 (also forward biased), and back down to the more negative side of the secondary. The negative alternation of the 12.6 V AC secondary appears across the load resistor R_L and on Channel 2 of the oscilloscope screen as shown in Figure 3-16(b). So, for the negative alternation of the transformer secondary, two diodes (D_2 and D_4) are forward biased and thus pass current. Notice that current doesn't go through D_1 or D_3 because now they are reverse biased and thus no current flows through them.

The bridge has provided full-wave rectification; it has changed the AC signal to a pulsating DC

signal. We consider the output signal pulsating DC because it still varies in amplitude like an AC waveform, but it has no negative alternation. The negative alternation from the 12.6 V AC input to the circuit has been changed to a positive waveform at the 12 kΩ load resistor because for both alternations the current goes through the load resistor in the same *direction*. Remember, one of the defining characteristics of AC is that it reverses direction.

Of course, in real life we can't freeze an AC waveform. Figure 3-17 shows the waveforms as you would see them on an o-scope. The top of Figure 3-17 shows the AC waveform going into the circuit, and the bottom shows the waveform that appears across the 12 kΩ load resistor.

This circuit is called a full-wave rectifier because both the positive alternation and negative alternation of the AC input signal are reproduced across the load resistor. Full-wave bridge rectifiers are used in many power supply applications. They are an improvement over the full-wave center-tapped rectifier because they do not require the use of a center-tapped transformer and provide more load voltage than the half-wave rectifier and the full-wave center-tapped rectifier. How much load voltage? you ask. Let's figure it out.

From Figure 3-14, we see that the transformer secondary RMS voltage is 12.6 V AC.

So, the peak voltage across the secondary,

$$V_P = \frac{V_{RMS}}{.707} = \frac{12.6\,V}{.707} = 17.82\,V$$

Now, for a bridge rectifier we need a new equation to determine the peak voltage across the load resistor. In a bridge rectifier, two diodes are conducting for each alternation. If each diode drops 0.6 V, then the total voltage drop is 0.6 V + 0.6 V = 1.2 V. This leads to a new equation.

EQUATION 3-3

Full-wave bridge rectifier (without filter capacitor)

V_P (across the load) = V_P (secondary) − 1.2 V

V_P = peak voltage

So, for Figure 3-14, V_P (across the load) = 17.82 V − 1.2 V = 16.62 V

Once we have the peak voltage across the load, we can calculate the average value; that is, the pulsating DC voltage. The equation for the average voltage in a full-wave bridge rectifier is the same used for a full-wave center-tapped rectifier.

(Equation 3-2)

$$V_{AVG} = .637 \ (V_P \text{ across the load})$$

$$V_{AVG} = .637 \ (V_P \text{ across the load})$$
$$= .637 \ (16.62 \text{ V}) = 10.58 \text{ V}$$

Thus, for Figure 3-14, the average (pulsating DC) voltage across the 12 kΩ load resistor is 10.58 V. This is twice the voltage that appears across the load for a full-wave center-tapped rectifier with the same transformer secondary voltage. (See calculations for Figure 3-8 to compare results.) Because they provide more load voltage than half-wave or full-wave center-tapped rectifiers, bridge rectifiers are the preferred rectifiers for power supplies.

So, 10.58 V is what you should measure with your DMM set to the DC voltage setting. It may be a little lower or higher since we estimated the diode voltage drops @ 0.6 V. Remember, once an AC signal is applied to a rectifier circuit, it is no longer AC—it is pulsating *DC*.

Let's do another example to reinforce the calculations we've just learned.

For Figure 3-18, you should measure @ 20.85 V across the load resistor R_L if you use your DMM set to the DC voltage setting. It may be a little lower or higher since we estimated the diode voltage drops @ 0.6 V.

Let's build a live version of the full-wave bridge rectifier.

EXAMPLE 2

Situation

Determine V_{AVG} for the circuit of Figure 3-18.

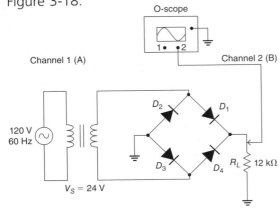

Solution

For Figure 3-18,

$$V_P(\text{across secondary}) = \frac{V_{RMS}}{.707} = \frac{24 \text{ V}}{.707}$$

$$= 33.94 \text{ V}$$

$$V_P(\text{across the load}) = 33.94 \text{ V} − 1.2 \text{ V}$$

$$= 32.74 \text{ V}$$

$$V_{AVG} = .637 \ (V_P) = .637 \ (32.74 \text{ V})$$

$$= 20.85 \text{ V}$$

FIGURE 3-18 Full-wave bridge rectifier circuit (© Cengage Learning 2012)

Full-Wave Bridge Rectifier

Materials, Equipment, and Parts:

- Digital Multimeter (DMM) with test leads.

- Single trace oscilloscope w/BNC-to-alligator leads.

- 12.6 V AC power supply.

- Breadboard and connecting wires.

- 1N4001 diode or equivalent (4).

- 1 μF electrolytic capacitor.

- 100 μF electrolytic capacitor.

- 12 kΩ fixed resistor.

Discussion Summary:

A full-wave bridge rectifier uses four diodes in a diamond-shaped configuration to convert AC to pulsating DC. On the positive alternation of the AC source, two diodes are forward biased, current flows through the diodes and load resistor, and the positive alternation appears across the load resistor R_L. On the negative alternation of the AC source, the other two diodes are forward biased, current flows through the diodes and load resistor, and the negative alternation appears as a positive voltage across the load resistor R_L.

Procedure:

SAFETY FIRST. Eye protection should always be worn when working with live voltages. Before powering on a live circuit, always check with your instructor.

 Build the circuit shown in Figure 3-19. We will not use Channel 1 of the o-scope. Connect the red lead of Channel 2, or B, of the o-scope to the top of the 12 kΩ load resistor. Connect the black lead of Channel 2, or B, a wire from the left side of the diamond, and a wire from the bottom of the load resistor to one point (hole) of the breadboard.

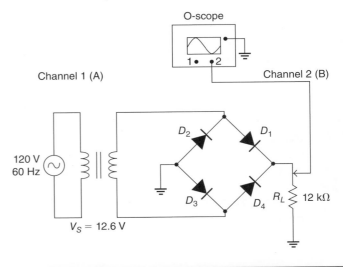

FIGURE 3-19 Full-wave bridge rectifier circuit (© Cengage Learning 2012)

(continues)

LAB ACTIVITY 3-3

(continued)

2   Set your DMM to measure AC voltage. Use your DMM to measure and record the AC power supply voltage (@ 12.6 V).

$V_S =$ _____

3   Use the following AC equation to determine the peak voltage across the transformer secondary:

V_P (across secondary) $= \dfrac{V_{RMS}}{.707} =$

V_P (across secondary) $=$ _____

4   Use Equation 3-3 to determine the peak voltage across the load resistor.

(***Equation 3-3***) V_P (across the load) $= V_P$ (secondary) $- 1.2$ V

V_P (across the load) $=$ _____

5   Use Equation 3-2 to calculate the average or pulsating DC voltage across the load.

(***Equation 3-2***) $V_{AVG} = .637$ (V_P across the load)

$V_{AVG} =$ _____

6   Have your instructor check your circuit.

7   Power on the circuit. Set your Mode switch to Channel 2. Adjust your oscilloscope volts/division settings and time/division settings to get the output waveform to appear like that of Channel 2 of Figure 3-17.

8   Set your DMM to measure DC voltage. Remember, the output voltage across the load resistor is now *pulsating DC*, so you have to set your meter to measure DC voltage. Measure and record the average (pulsating DC) voltage across the 12 kΩ load resistor.

$V_{AVG} =$ _____

9   Compare the calculated average voltage across the load resistor (Step 5) to the measured average voltage across the load resistor (Step 8). Explain any differences.

10   Power off the circuit.

11   *Observing polarity*, carefully insert a 1 μF electrolytic capacitor in parallel with and to the left of the 12 kΩ load resistor.

LAB ACTIVITY 3-3

12 Have the instructor check your circuit.

13 Power on. What happened to the waveform across the 12 kΩ R_L? Did the capacitor provide filtering?

14 Power off the circuit.

15 *Observing polarity*, carefully replace the 1 μF electrolytic capacitor with the 100 μF electrolytic capacitor.

16 Have the instructor check your circuit.

17 Power on. What happened to the waveform across the 12 kΩ R_L? Did the capacitor provide filtering? How does the voltage waveform using the 100 μF capacitor compare to the waveform using the 1 μF capacitor from Step 11?

18 Power off the circuit and have the instructor check your results.

19 Build the circuit on Multisim®, Electronics Workbench®, or SPICE and then compare the results.

In this section, we learned that the full-wave bridge rectifier has four diodes connected in a diamond shape. In real-life applications, especially low-end (inexpensive) ones, you may find a bridge rectifier with four diodes, although the diodes may be mounted side by side. It's more common, however, for manufacturers to construct the four diodes within a single package. Figure 3-20(a) shows several examples of bridge rectifier packages, and Figure 3-20(b) shows the schematic symbol for a bridge rectifier.

A bridge rectifier package contains four diodes, and often the connections are marked on the package. The schematic symbol of Figure 3-20(b) also shows the connections for the bridge rectifier. The AC inputs connect to the terminals of the IC that are labeled with sine waves; the terminal with a negative sign (−) connects to the negative side of the load resistor; and the terminal with a positive sign (+) connects to the positive end of the load resistor.

The bridge rectifier on the right of Figure 3-20(a) is an integrated circuit. We discussed integrated circuits briefly in Chapter 2 when we performed Lab Activity 2-3 using an optoisolator. Since the bridge rectifier is the second semiconductor device we've encountered that could be manufactured in IC form, it's time to discuss ICs in greater detail.

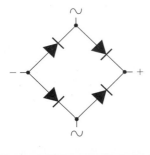

a)

Courtesy of Tracy Grace Leleux

b)
© Cengage Learning 2012

FIGURE 3-20   Bridge rectifier: a) Examples of bridge rectifier packages and b) Schematic symbol

3-5 INTEGRATED CIRCUITS (ICs)

The first integrated circuit, or IC, was co-invented by Jack Kilby and Robert Noyce in 1958. These men were working independently, in different parts of the country, and for different companies (Kilby for Texas Instruments, Noyce for Fairchild Semiconductor). Separately, they each invented circuitry that became the forerunner for the IC, so they are considered co-invented.

Remember, an integrated circuit, or "IC" or "chip," contains several electronic devices housed in one package—one package, many devices. This distinguishes an IC from other electronic devices. A **discrete device**—such as a resistor, capacitor, or diode—is one package, one device. ICs have revolutionized the electronics and computer industries because they require less space than discrete devices. For example, the 741 op amp IC is about the size of a dime and can contain the equivalent of forty electronic devices, including transistors, resistors, diodes, and capacitors. Before the IC, a 741 op amp built from discrete devices would take up the space of a slice of bread. Integrated circuits also consume less power than their discrete counterparts.

An IC can contain any combination of diodes, resistors, capacitors, transistors (to be discussed in Chapter 4), digital logic gates, etc. Keep in mind, however, that manufacturers don't just lump together a bunch of different devices into one housing: the manufacturing process is quite complicated and involved. For example, you can't smash a bridge rectifier IC with a hammer and expect to find four little diodes. As technicians, we're more concerned with the final product before us.

INTERNET ALERT

Check the website http://www.youtube.com/watch?v=i8kxymmjdoM to discover more about the IC manufacturing process.

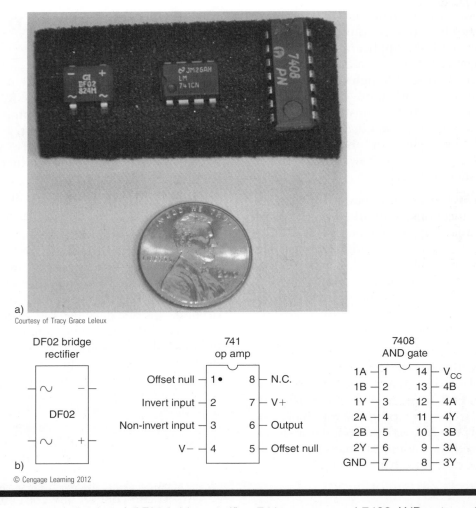

© Cengage Learning 2012

FIGURE 3-21   Integrated circuits: a) DF02 bridge rectifier, 741 op amp, and 7408 AND gate and b) Pin configurations

ICs come in many sizes and pin configurations. The most common ICs range from the size of a dime to a stick of Dentyne gum and have four, six, eight, or fourteen pins. Figure 3-21 shows three common ICs and their pin configurations.

Manufacturers of ICs often put a dot or a notch on an IC for identification purposes. If there is a dot, then the pin closest to the dot is pin 1 as shown in the 741 op amp of Figure 3-21(b).

When building a circuit with an IC, if the notch is up or at the twelve o'clock position, pin 1 is the first pin on the left as shown in both the 741 op amp and the 7408 AND gate in Figure 3-21(b). The remaining pins are numbered in counter-clockwise sequence. (The 741 op amp IC is unusual in that it has both a dot *and* a notch.)

Other than locating pin 1, however, there's little you can tell about an IC by its outer case. Like discrete diodes and transistors, ICs have manufacturer part numbers, and the only way to tell the pin configuration and function of an IC is to look up its part number in the *NTE* catalog or another semiconductor replacement guide. *Look it up to hook it up.*

For example, the DF02 bridge rectifier of Figure 3-21 has no dot or notch, but it has the input pins labeled on the housing. The two pins labeled with sine waves connect to the AC inputs from the source voltage; the pin with a negative sign (−) connects to the negative side of the load resistor, and the pin with a positive sign (+) connects to the positive end of the load resistor. If an experienced electronic technician saw this IC, she would understand the labeling and would recognize that the IC is a bridge rectifier. However, if she needed more information such as voltage, current, or power limits, she would have to *Look it up to hook it up.*

Some IC pin diagrams show the internal circuitry, like the 4N25 optoisolator back in Chapter 2, but most diagrams just label the pins. Some pins have no internal connection to a device and are marked N.C., for "no connection," like pin 8 of the 741 op amp in Figure 3-21(b). The 7408 AND gate shown in Figure 3-21 is a typical IC you will study later in digital class. Manufacturers of digital ICs have a specific scheme for identifying IC part numbers.

The IC is a sealed unit, so as a technician, you have access only to the external pins. If one internal device burns out, you have to replace the entire IC. In this regard, an IC is like the seven-segment display.

The three ICs in Figure 3-21 are mounted on printed circuit boards in a process called **through-hole technology**. In through-hole technology, the pins or leads of the IC are inserted directly into pre-drilled holes in a circuit board and then soldered on the opposite side to circular pads. Sometimes the ICs are mounted in sockets, and the socket leads are inserted into the holes and then soldered. Circuit boards using through-hole technology can be assembled manually or by robots, and they are usually easy to solder and de-solder when replacing a bad IC. In addition to ICs, many other electronic devices such as resistors, capacitors, and coils are mounted to a circuit board using through-hole technology.

Surface Mount Technology (SMT) involves the soldering of ICs directly onto the surface or to one side of a printed circuit board. Surface Mount Technology was developed in the 1960s and became commercially widespread in the 1980s. Surface Mount Technology is cheaper because it doesn't require predrilled holes and because more ICs can be fitted into a smaller space. Also, surface mount devices use less power. Many expansion cards for computer—such as network cards and sound cards—are built using surface mount technology.

The main drawback to surface mount technology is the fact that the leads are so small and closely spaced that it's hard to solder and de-solder the ICs. In fact, most SMT circuit boards are manufactured by robots. Resistors, capacitors, and diodes are also mounted using surface mount technology, and they can be as small as the head of a pencil. For commercial applications, many printed circuit boards are manufactured using a combination of both surface mount technology and through-hole technology.

The schematic designator for ICs begins with the letter *U* followed by a number that is usually in subscript form. For example, if you have three ICs in a circuit, they would be labeled U_1, U_2, and U_3. Some manufacturers label ICs as IC1, IC2, and IC3, with or without the subscript.

3-6 A COMPLETE POWER SUPPLY

Now that we've studied the individual stages of a power supply, it's time to see the way all the stages and their components work together. Figure 3-22 shows a complete, though simple, power supply. The power supply has a transformer with a 12.6 V AC secondary, a bridge rectifier, an electrolytic filter capacitor, a Zener diode, and a 10 kΩ load resistor. It's not likely that you will see this design in any commercial application, but it demonstrates what happens in a typical power supply.

The analysis of the circuit uses a combination of equations and reasoning we learned in this chapter.

1) The transformer reduces or steps down the incoming 120 V AC to 12.6 V AC. So, the first value we need is the peak voltage across the 12.6 V AC transformer secondary winding.

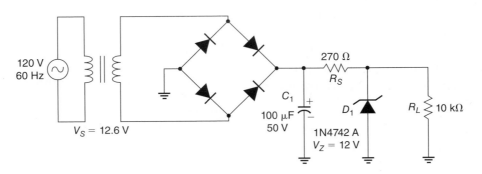

FIGURE 3-22   Complete power supply (© Cengage Learning 2012)

$$V_P \text{ (across secondary)} = \frac{V_{RMS}}{.707} = \frac{12.6\,V}{.707} = 17.82\,V$$

2) Now you may be tempted to use Equation 3-3, which we used for the full-wave bridge rectifier without a filter capacitor. However, with a filter capacitor, the equation goes out the window. We used filter capacitors in Lab Activities 3-1, 3-2, and 3-3 to see the way a capacitor reduces the pulsating action of a rectified wave, but we didn't measure the voltage across the filter capacitor. For most labs, a 100 µF electrolytic capacitor will provide excellent filtering and still maintain a high output voltage across the load. This is because the filter capacitor C_1 (really any capacitor) charges to the *peak* value across the secondary winding minus the diode drops on either alternation.

Link to Prior Learning

A capacitor always charges to the peak voltage in any circuit, including when it's used to filter the output of a rectifier.

$$V_P \text{ (across cap)} = V_P \text{ (across secondary)}$$
$$- 1.2\,V = 17.82\,V - 1.2\,V = 16.62\,V$$

The filter capacitor charges rapidly but discharges slowly, so it also smoothes the waveform by removing the pulsating action.

3) This 16.62 V_P would normally appear across the 10 kΩ load resistor. However, the Zener diode has a V_Z of 12 V, so the Zener turns on, diverts current away from the 10 kΩ load resistor, and maintains a steady voltage of @ 12 V DC across the load resistor. In commercial applications, a voltage regulator IC might be used instead of just one Zener diode.

4) The input to the power supply is 120 V AC, and the output is a filtered and regulated 12 V DC. This process occurs in clock radios, stereo receivers, laptop computers, and almost every system that is connected to an AC outlet.

Let's discuss one last point about power supplies. The output voltage of a power supply shouldn't vary much under changing load conditions such as an increase in current. For example, many people run gas-powered generators during blackouts. When a generator is first started and no appliance is connected to it, the generator makes a steady sound; this is called the "no-load" condition, and the output voltage for each outlet should be @ 120 V. When an appliance is connected to the generator (the "full-load" condition), you can hear the generator change to a lower pitch, or "dog out," which will cause the output voltage to drop slightly. This change in output voltage from a no-load to a full-load condition is called the **voltage regulation** for the power supply, and it can be calculated using Equation 3-4. The less the change or % of voltage regulation, the better the power supply.

EQUATION 3-4

Power supply voltage regulation

$$\% \text{ regulation} = \frac{V_{NL} - V_{FL}}{V_{FL}}(100)$$

V_{NL} = output voltage under no-load conditions

V_{FL} = output voltage under full-load conditions

For example, say you have a power supply that measures 12 V without a load attached and measures 11.5 V with a load attached and drawing current. What is the % regulation?

$$\% \text{ regulation} = \frac{V_{NL} - V_{FL}}{V_{FL}}(100)$$
$$= \frac{12\,V - 11.5\,V}{11.5\,V}(100) = \frac{.5\,V}{11.5\,V}(100)$$
$$= 4.34\%$$

The voltage regulation for this power supply—4.34%—is very good. Remember, the lower the % of voltage regulation, the better the system. When buying a power supply or generator, the owner's manual should list the % regulation.

Let's do another example to reinforce the calculation we've just learned.

EXAMPLE 3

Situation

A power supply measures 120 V without a load attached and measures 112 V with a load attached and drawing current. What is the % regulation?

Solution

$$\% \text{ regulation} = \frac{V_{NL} - V_{FL}}{V_{FL}}(100)$$

$$= \frac{120\text{ V} - 112\text{ V}}{112\text{ V}}(100)$$

$$= \frac{8\text{ V}}{112\text{ V}}(100)$$

$$= 7.14\%$$

Troubleshooting Tip

The first step in troubleshooting is to do a visual inspection. Power supply problems typically involve either 0 voltage output or low voltage output. For 0 voltage output, first check the fuse. Fuses naturally wear out after long use. If you replace the fuse (with one with the same voltage and current rating) and the power supply power light goes out again, there's a _short_ somewhere, either in the transformer or in a diode in the rectifier circuit. If a power supply has low voltage output, there's probably an _open_ somewhere, either in the transformer or in a diode in the rectifier circuit. Use your DMM to check these components.

The voltage regulation for this power supply—7.14%—is very good. Remember, the lower the % of voltage regulation, the better the system. When buying a power supply or generator, the owner's manual should list the % regulation.

Now that we've analyzed a power supply in detail, it's time to build a complete yet simple power supply on the breadboard.

LAB ACTIVITY 3-4

A Complete Power Supply

Materials, Equipment, and Parts:

- *NTE* catalog or Internet access, www.nteinc.com

- Digital Multimeter (DMM) with test leads.

- 12.6 V AC power supply.

- Breadboard and connecting wires.

- 1N4001 diode (4) or IC bridge rectifier such as DB102 (NTE 5332), DF02 (NTE 5332), or one that can handle 1 A for safety sake.

- 1N4742A Zener diode (12 V_Z), or close to 12-V value.

- 100 µF electrolytic capacitor.

- 270 Ω fixed resistor.

- 10 kΩ fixed resistor.

Discussion Summary:

A complete power supply has an AC input voltage and provides a filtered and regulated DC output voltage across the load. The different stages are voltage reduction, rectification, filtering, and regulation. The Voltage Reduction stage uses a transformer; the Rectification stage uses combinations of diodes; the Filtering stage uses an electrolytic capacitor; and the Regulation stage uses a Zener diode.

Procedure:

SAFETY FIRST. Eye protection should always be worn when working with live voltages. Before powering on a live circuit, always check with your instructor.

1 Build the circuit shown in Figure 3-23. Connect the grounded end of the bridge rectifier, the negative side of the electrolytic capacitor, the anode of the Zener diode, and the negative side of the load resistor to one point (hole) of the breadboard. Ensure the 100 µF electrolytic capacitor is connected correctly. Have your instructor check your circuit.

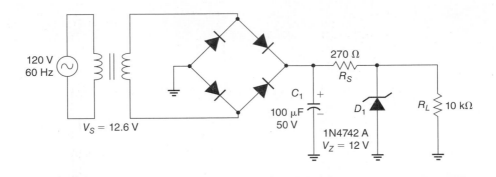

FIGURE 3-23 Complete power supply (© Cengage Learning 2012)

(continues)

LAB ACTIVITY 3-4

(continued)

2 Set your DMM to measure AC voltage. Power on your circuit and use your DMM to measure and record the AC power supply voltage (@ 12.6 V).

$V_S =$ _____

3 Power off. Use the following AC equation to determine the peak voltage across the transformer secondary:

V_P (across secondary) $= \dfrac{V_{RMS}}{.707} =$

V_P (across secondary) $=$ _____

4 Use the following equation to calculate the DC voltage across filter capacitor C_1.

V_P (across cap) $= V_P$ (across secondary) $- 1.2$ V

V_P (across cap) $=$ _____

5 Set your DMM to measure DC voltage. Power on and measure and record the DC voltage across the capacitor.

$V_{C1} =$ _____

6 Measure the DC voltage across the 10 kΩ load resistor.

$V_{RL} =$ _____

7 Why is the measured DC voltage across the 10 kΩ load resistor in Step 6 different than the measured DC voltage across capacitor C_1 in Step 5?

8 Power off the circuit and have the instructor check your results.

9 Build the circuit on Multisim®, Electronics Workbench®, or SPICE and then compare the results.

3-7 VOLTAGE MULTIPLIERS (DOUBLERS AND TRIPLERS)

The diodes we've studied so far have been used for various applications. The PN junction diode is used as a rectifier; the LED as a panel indicator light; and the Zener diode as a voltage regulator.

Diodes, along with capacitors, also can be used as voltage multipliers. A **voltage multiplier** is used to change AC to pulsating DC *and* increase the output voltage of a circuit. Voltage multipliers are used in low-power circuits where transformers are impractical or too costly. Although step-up transformers can be used to increase the output voltage of a circuit, transformers can't change AC to DC.

Voltage multipliers are also used in high-voltage DC applications, including cathode ray tubes for TVs, copy machines, and bug zappers. Heinrich Greinacher, a Swiss physicist, invented the first vacuum tube multiplier in 1913. Most voltage multipliers today use semiconductors.

Multiplier circuits include voltage doublers and voltage triplers. Figure 3-24 shows a voltage doubler circuit.

A **voltage doubler** circuit uses two diodes and three capacitors to double the peak output voltage of an AC voltage source while changing it to pulsating DC. In Figure 3-24, diode D_1

and capacitor C_1 work together as a unit, and diode D_2 and capacitor C_2 work as a unit. Capacitor C_3 is a filter capacitor that combines the voltage from C_1 and C_2. The voltage across C_3 will also appear across the 82 kΩ load resistor because the two components are connected in parallel. Resistor R_S limits the current through both diodes.

For learning purposes only, we are going to "freeze" the AC voltage source. When the 6.3 V AC source is positive on the top and negative on the bottom as shown in Figure 3-25, diode D_1 is forward biased—positive to anode and negative to cathode—and current flows through R_S, through D_1, and charges capacitor C_1 to @ 8.31 V (the 8.91 V peak voltage minus the 0.6-V diode voltage drop), which is now pulsating DC. At this point, D_2 is reverse biased, so no current flows through it.

When the 6.3 V AC source is negative on the top and positive on the bottom as shown in Figure 3-26, diode D_2 is forward biased—positive to anode and negative to cathode—and current flows through D_2 and charges capacitor C_2 to @ 8.31 V (the 8.91 V peak voltage minus the 0.6-V diode voltage drop), which is now pulsating DC. At this point, D_1 is reverse biased, so no current flows through it.

At this brief moment, both C_1 and C_2 are each charged to @ 8.31 V. Since C_3 is in parallel with the *combination* of C_1 and C_2, C_3 will be charged to the sum of these voltages, or @ 16.62 V. Also,

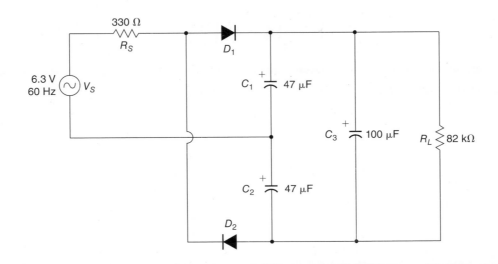

FIGURE 3-24 Voltage doubler circuit (© Cengage Learning 2012)

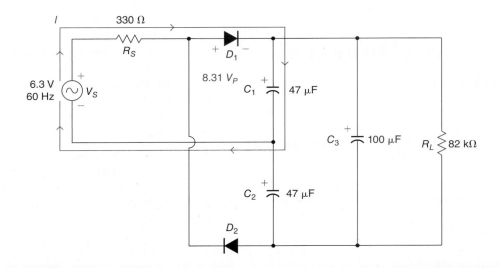

FIGURE 3-25   Voltage doubler circuit with diode D_1 conducting (© Cengage Learning 2012)

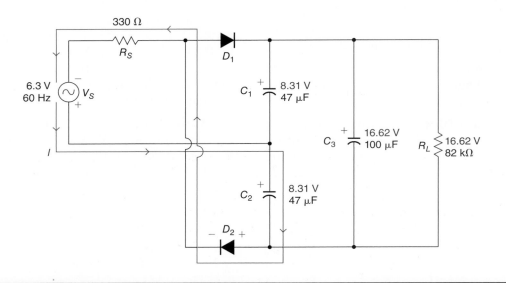

FIGURE 3-26   Voltage doubler circuit with diode D_2 conducting (© Cengage Learning 2012)

C_3 will filter the pulsating DC from the C_1-C_2 combination. Since C_3 is in parallel with the 82 kΩ load resistor, the 16.62 V across C_3 will also appear across the 82 kΩ load resistor. Of course, the actual DC voltage measured across the 82 kΩ load resistor will be slightly less or more because we estimated the diode voltage drops @ 0.6 V.

The voltage doubler has roughly doubled the source peak voltage, providing rectified and filtered DC to the load resistor. This circuit is called a full-wave voltage doubler because both alternations of the AC voltage source help produce a voltage across the load. Let's build it live.

LAB ACTIVITY **3-5**

Voltage Doubler

Materials, Equipment, and Parts:

- Digital Multimeter (DMM) with test leads.

- 6.3 V AC power supply.

- Breadboard and connecting wires.

- 1N4001 diode (2).

- 47 µF electrolytic capacitor (2).

- 100 µF electrolytic capacitor.

- 330 Ω fixed resistor.

- 82 kΩ fixed resistor.

Discussion Summary:

A voltage doubler uses diodes and capacitors to double the peak output of a source voltage, while at the same time providing rectified and filtered DC to the load resistor. The circuit we'll build is called a full-wave voltage doubler because both alternations of the AC voltage source help produce a voltage across the load.

Procedure:

SAFETY FIRST. Eye protection should always be worn when working with live voltages. Before powering on a live circuit, always check with your instructor.

1 Build the circuit shown in Figure 3-27. Ensure the 47 µF and 100 µF electrolytic capacitors are connected correctly.

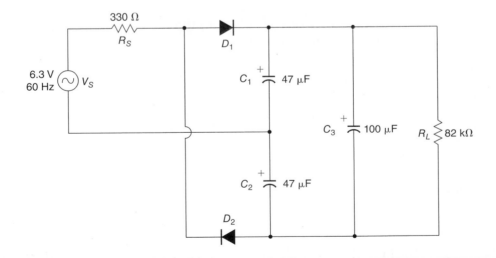

FIGURE 3-27 Voltage doubler circuit (© Cengage Learning 2012)

Figure 3-28 shows a breadboard version of the voltage doubler of Figure 3-27. Have your instructor check your circuit.

(continues)

LAB ACTIVITY 3-5

(continued)

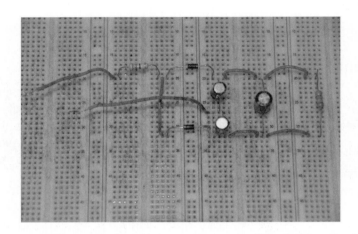

FIGURE 3-28   Voltage doubler breadboard circuit (Courtesy of Tracy Grace Leleux)

2   Set your DMM to measure AC voltage. Power on your circuit and use your DMM to measure and record the AC power supply voltage (@ 6.3 V).

$V_S =$ _____

3   Power off. Use the following AC equation to determine the peak voltage of the AC power supply:

$$V_P \text{ (AC power supply)} = \frac{V_{RMS}}{.707} =$$

V_P (AC power supply) = _____

4   The peak voltage across either C_1 or C_2 is equal to the AC power supply peak voltage minus the 0.6-V diode voltage drop across D_1 or D_2.

V_P (across C_1 or C_2) = V_P (AC power supply) $-$ .6 V =

V_P (C_1 or C_2) = _____

5   Set your DMM to measure DC voltage. Power on and measure and record the DC voltage across capacitor C_3.

$V_{C3} =$ _____

6   Measure the DC voltage across the 82 kΩ load resistor.

$V_{RL} =$ _____

7   Is the measured DC voltage of Step 5 and Step 6 roughly double the voltage of Step 4?

8   Power off the circuit and have the instructor check your results.

9   Build the circuit on Multisim®, Electronics Workbench®, or SPICE and then compare the results.

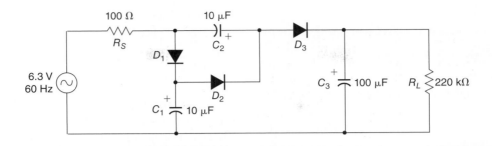

FIGURE 3-29 Voltage tripler circuit (© Cengage Learning 2012)

The voltage doubler isn't the only voltage multiplier circuit in use. Figure 3-29 shows a voltage tripler circuit.

A **voltage tripler** uses three diodes and three capacitors to triple the peak output voltage of an AC voltage source and change it to pulsating DC. In Figure 3-29, diode D_1 conducts to charge capacitor C_1, diode D_2 conducts to charge capacitor C_2, and diode D_3 conducts to charge capacitor C_3. Capacitor C_3 adds the voltages from C_1 and C_2 to its own voltage; C_3 also provides filtering. The voltage across C_3 will also appear across the 220 kΩ load resistor because they are connected in parallel.

In Figure 3-29, each capacitor charges to @ 8.31 V (the 8.91 V peak voltage minus the 0.6 V diode voltage drop), which is now pulsating DC. Capacitor C_3 will be charged to the sum of these voltages, or @ 24.93 V. Since C_3 is in parallel with the 220 kΩ load resistor, the 24.93 V across C_3 will also appear across the 220 kΩ load resistor. Of course, the actual DC voltage measured across the 220 kΩ load resistor will be slightly less or more because we estimated the diode voltage drops @ 0.6 V. The voltage tripler has roughly tripled the source peak voltage, providing rectified and filtered DC to the load resistor.

Voltage multipliers can be made to quadruple a source peak voltage, but low-power commercial applications rarely exist beyond the tripler circuit. Since energy can't be created out of nothing, a trade-off is made with voltage multiplier circuits. To get the voltage to increase, the current available to the load must decrease. Because the load current becomes so small in multiplier circuits beyond the voltage tripler configuration, quadrupler circuits are rarely used.

CHAPTER SUMMARY

Power supplies are used in many electronic systems to provide the voltages that drive the subsystems. Technicians need to identify and troubleshoot the electronic devices within a power supply. Diodes play an important role in power supplies along with other electronic devices. The stages of a power supply include the Voltage Reduction stage, where a transformer reduces or step downs the voltage; the Rectification stage, where PN junction diodes change the AC voltage to pulsating DC; the Filtering stage, where filter capacitors remove the pulsating portion of the waveform; and the Regulation stage, where a Zener diode provides a steady DC voltage to the load.

A half-wave rectifier uses a diode to convert AC to pulsating DC. On the positive alternation of the AC source, the diode is forward biased, current flows through the diode and load resistor, and the positive alternation appears across the load resistor R_L. On the negative alternation of the AC source, the diode is reverse biased, no current flows through the diode or load resistor, and no waveform appears across the load resistor.

A full-wave center-tapped rectifier uses two diodes to convert AC to pulsating DC. On the positive alternation of the AC source, one diode is forward biased, current flows through the diode and load resistor, and the positive alternation appears across the load resistor R_L. On the negative alternation of the AC source, the other diode is forward biased, current flows through the diode and load resistor, and the negative alternation appears as a positive voltage across the load resistor R_L.

A full-wave bridge rectifier uses four diodes in a diamond-shaped configuration to convert AC to pulsating DC. On the positive alternation of the AC source, two diodes are forward biased, current flows through the diodes and load resistor, and the positive alternation appears across the load resistor R_L. On the negative alternation of the AC source, the other two diodes are forward biased, current flows through the diodes and load resistor, and the negative alternation appears as a positive voltage across the load resistor R_L.

An integrated circuit, or IC or chip, contains several electronic devices housed in one package. This distinguishes an IC from other electronic devices. Conversely, a discrete device—such as a resistor, capacitor, or diode—is one device housed in one package. ICs have revolutionized the electronics and computer industries because they require less space than discrete devices. ICs also consume less power than their discrete counterparts.

Many ICs are mounted on printed circuit boards using through-hole technology, where the pins or leads of the IC are inserted directly into predrilled holes in a circuit board and then soldered on the opposite side to circular pads. Circuit boards using through-hole technology are easy to solder and de-solder when replacing a bad IC. In addition to ICs, many other electronic devices such as resistors, capacitors, and coils are mounted on a circuit board using through-hole technology.

Surface Mount Technology (SMT) involves the soldering of ICs directly onto the surface or to one side of a printed circuit board. Surface mount technology is cheaper because it doesn't require predrilled holes and more ICs can be fit in a smaller space. Also, surface mount devices use less power. The main drawback to SMT is that the leads are small and closely spaced, making it hard to solder and de-solder the ICs. In fact, most SMT circuit boards are manufactured by robots. Resistors, capacitors, and diodes are also mounted using surface mount technology, and they can be as small as the head of a pencil. For commercial applications, many printed circuit boards are manufactured using a combination of both surface mount technology and through-hole technology.

The output voltage of a power supply varies under changing load conditions such as an increase in current. The change in output voltage from a no-load to a full-load condition is called the voltage regulation for the power supply.

The smaller the voltage regulation, the better the power supply.

Diodes, along with capacitors, also can be used as voltage multipliers. A voltage multiplier is used to both change AC to pulsating DC and increase the voltage output. Voltage multipliers are used in low-power circuits where transformers are impractical or too costly. Multiplier circuits include voltage doublers and voltage triplers. A voltage doubler uses two diodes and three capacitors to double the peak output of a source voltage while at the same time providing rectified and filtered DC to the load resistor. A voltage tripler uses three diodes and three capacitors to triple the peak output voltage of an AC voltage source.

CHAPTER EQUATIONS

Full-wave center-tapped rectifier (without filter capacitor)

(Equation 3-1)
$$V_P \text{ (across the load)} = \frac{V_P \text{ (secondary)}}{2} - .6 \text{ V}$$
V_P = peak voltage

Full-wave rectifier (without filter capacitor)

(Equation 3-2)
$$V_{AVG} = .637 (V_P)$$
V_{AVG} = pulsating DC voltage across the load
V_P = peak voltage across the load

Full-wave bridge rectifier (without filter capacitor)

(Equation 3-3)
$$V_P \text{ (across the load)} = V_P \text{ (secondary)} - 1.2 \text{ V}$$
V_P = peak voltage

Power supply voltage regulation

(Equation 3-4)
$$\% \text{ regulation} = \frac{V_{NL} - V_{FL}}{V_{FL}}(100)$$
V_{NL} = output voltage under no-load conditions
V_{FL} = output voltage under full-load conditions

CHAPTER REVIEW QUESTIONS

Chapter 3-1

1. Draw and label the four stages of a power supply, including the electronic device used for each stage.

Chapter 3-2

2. Explain the difference between conventional current flow and electron flow.

3. When does current flow through the load in a half-wave rectifier circuit?

4. The circuit of Figure 3-30 is a positive half-wave rectifier. What would you have to do to the circuit to make it a negative half-wave rectifier?

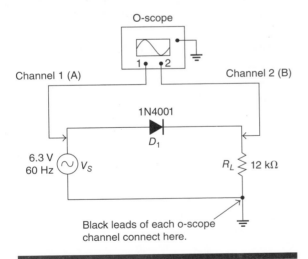

FIGURE 3-30 Half-wave rectifier circuit (© Cengage Learning 2012)

5. Explain how an electrolytic capacitor provides filtering in a power supply.

6. What precaution must one take when replacing an electrolytic capacitor?

7. Why aren't half-wave rectifiers commonly used in commercial power supplies?

Chapter 3-3

8. Calculate the average (pulsating DC) load voltage for the circuit of Figure 3-31.

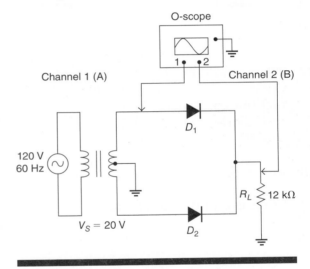

FIGURE 3-31 Full-wave center-tapped rectifier (© Cengage Learning 2012)

9. Sketch the output waveform for the circuit of Figure 3-31.

10. Name one disadvantage of a full-wave center-tapped rectifier.

Chapter 3-4

11. Calculate the average (pulsating DC) load voltage for the circuit of Figure 3-32.

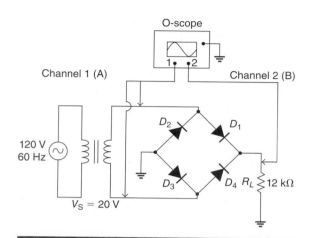

FIGURE 3-32 Full-wave bridge rectifier circuit (© Cengage Learning 2012)

12. What would be the advantage of a bridge rectifier IC versus a bridge rectifier constructed with four diodes in a diamond shape?

Chapter 3-5

13. What is an integrated circuit?

14. What is a discrete circuit?

15. Name one advantage and one disadvantage of an IC made with surface mount technology.

16. What letter is usually used as a schematic designator for ICs?

Chapter 3-6

17. Is the transformer in Figure 3-33 a step-up or step-down transformer?

18. Calculate the average (pulsating DC) load voltage for the circuit of Figure 3-33.

19. What is the purpose of resistor R_S in Figure 3-33?

20. A power supply has a no-load voltage of 120 V and a full-load voltage of 110 V. What is the power supply's % voltage regulation?

21. A power supply has a no-load voltage of 15 V and a full-load voltage of 13.2 V. What is the power supply's % voltage regulation?

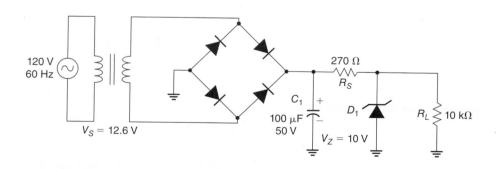

FIGURE 3-33 Complete power supply (© Cengage Learning 2012)

Chapter 3-7

22. **What is the approximate peak load voltage for the circuit of Figure 3-34?**

23. **In Figure 3-34, what is the purpose of capacitor C_3?**

24. **What is the approximate peak load voltage for the circuit of Figure 3-35?**

25. **Why aren't voltage multipliers used widely in high-current applications?**

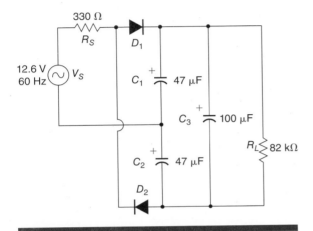

FIGURE 3-34   Voltage doubler circuit
(© Cengage Learning 2012)

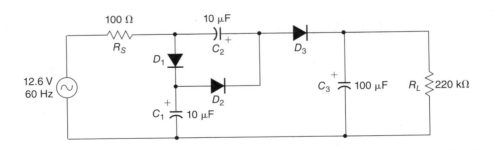

FIGURE 3-35   Voltage tripler circuit (© Cengage Learning 2012)

Transistors

Bipolar Junction Transistors (BJTs)

OBJECTIVES   *Upon completion of this chapter, you should be able to:*

- List the two applications for a transistor.
- Draw the three common transistor outlines or packages.
- Explain the difference between a Bipolar Junction Transistor (BJT) and a Field Effect Transistor (FET).
- Draw and label the schematic symbol for an NPN and a PNP transistor.
- Use a semiconductor replacement guide to determine critical information about a transistor.

- Identify a transistor's emitter, base, and collector terminals or leads.
- Use a DMM to test a transistor.
- Draw the correct biasing for an NPN transistor and explain the three types of current flow.
- Calculate the emitter current (I_E) in a BJT.
- Calculate the DC current gain (β) of a transistor.
- Build a circuit that uses a transistor as a high-speed switch.

MATERIALS, EQUIPMENT, AND PARTS

Materials, equipment, and parts needed for the lab experiments in this chapter are listed below:

- *NTE* catalog or Internet access, www.nteinc.com
- PC w/Multisim®, Electronics Workbench®, or SPICE.
- PC w/Solid State Challenges™.
- Digital Multimeter (DMM) with test leads.
- 5 V DC voltage source.
- 12 V DC voltage source.

- Variable DC voltage source, 0 V to 15 V.
- Function generator, 1 Hz to 60 Hz.
- Breadboard and connecting wires.
- 2N3904 or equivalent NPN transistor.
- One BJT with case style TO92 such as 2N3904 or 2N3906.

- One BJT with case style TO39 (top hat) such as 2N2219 or 2N3053.
- One BJT with case style TO3 or TO66 (spaceship) such as 2N3055 or 2N6373.
- Two-terminal LED (any color).
- 270 Ω fixed resistor.
- 1 kΩ fixed resistor (2).

GLOSSARY OF TERMS

Transistor A three-terminal semiconductor device used as a high-speed switch or as an amplifier

Bipolar Junction Transistor (BJT) A three-terminal current-controlled semiconductor device used as a high-speed switch or as an amplifier

Pin configuration The scheme for identifying the terminals or leads of many electronic devices; it is also called a pin layout or pin-out

DC current gain The current gain of a bipolar junction transistor that is the result of the collector current divided by the base current; it is also called beta (β) or h_{FE}

h_{FE} The DC current gain of a bipolar junction transistor that is the result of the collector current divided by the base current; also called beta (β)

Beta The DC current gain of a bipolar junction transistor that is the result of the collector current divided by the base current; it is also called h_{FE}

4-1 TRANSISTOR BASICS

The transistor was invented in late 1947 by John Bardeen, Walter Brattain, and William Shockley at Bell Labs in Murray Hill, New Jersey. This milestone occurred just two years after the end of World War II, and the transistor revolutionized electronics, spawning the solid-state or semiconductor industry.

As mentioned in Chapter 1, the transistor eventually replaced the vacuum tube in industrial and consumer electronics. The transistor is smaller, more efficient, and more rugged than its vacuum tube counterpart, the triode. The transistor and solid-state electronics led to the miniaturization of electronic circuitry. For example, Intel's first Pentium CPU introduced in 1993 had roughly 3.1 million transistors. By 2009, Intel's Xeon CPU had 781 million transistors! This miniaturization is still happening today.

INTERNET ALERT

Check the website http://www.porticus.org/bell/belllabs_transistor.html to discover more about the history of the transistor.

The **transistor** is a three-terminal semiconductor device that is used as a high-speed switch or as an amplifier. Transistors are designed for countless applications and come in many sizes and shapes. Figure 4-1 shows three common transistor outlines or packages that you may encounter as an electronic technician.

The two major categories of transistors are Bipolar Junction Transistors (BJTs) and Field Effect Transistors (FETs). BJTs and FETs differ in internal construction and method of operation. BJTs are current-controlled devices whose operation depends on the interaction of electrons *and* holes. FETs are voltage-controlled devices whose operation depends on the movement of either electrons *or* holes. BJTs are covered in this chapter and FETs in a later chapter. The transistors in Figure 4-1 are BJTs, but FETs can look the same. Again, the only way to tell the difference is *Look it up to hook it up*.

The **Bipolar Junction Transistor (BJT)** is a three-terminal device similar in composition to the

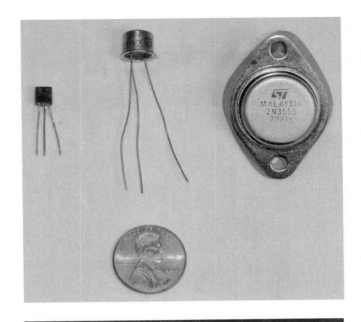

FIGURE 4-1 Three common transistor outlines or packages: a) TO92, b) TO39 top hat, and c) TO3 spaceship (Courtesy of Tracy Grace Leleux)

PN junction diode. Like the PN junction diode, the transistor is composed of P-type material such as boron and N-type material such as arsenic that are chemically combined in a silicon (Si) or germanium (Ge) base. The difference lies in the distribution of the P-type material and the N-type material.

BJTs are constructed in one of two ways: NPN or PNP. Figure 4-2 shows simplified drawings of both transistor constructions.

Figure 4-2(a) shows the NPN transistor, the one used in most electronic circuits. Notice that the NPN transistor has two parts N-type material, which is usually arsenic with excess electrons, and one part P-type material, which is usually boron with excess holes. (The PN junction diode discussed in Chapter 1 has one part P-type material and one part N-type material.)

The N-type material that has a large number of excess electrons is called the emitter. The P-type material with a very small number of excess holes is called the base. The N-type material with a small number of excess electrons is called the collector.

Figure 4-2(b) shows a PNP transistor. The PNP transistor has two parts P-type material and one part N-type material. The emitter region in this case has a large number of excess holes, the base region has a very small number of excess electrons, and the collector region has a small number of excess holes. Because PNP and NPN transistors differ in both

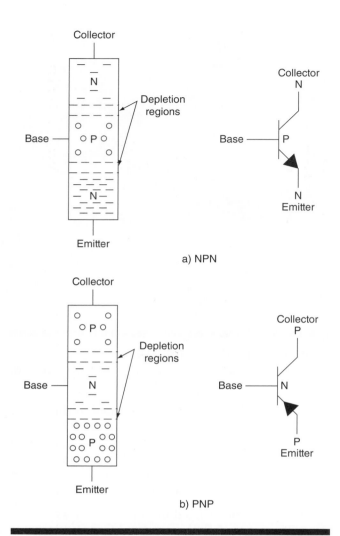

FIGURE 4-2   Transistor constructions and schematic symbols: a) NPN transistor and b) PNP transistor (© Cengage Learning 2012)

construction and operation, you can't replace a PNP transistor with an NPN transistor, or vice versa. *Always replace an NPN transistor with an NPN transistor, and always replace a PNP transistor with a PNP transistor.*

Notice in both schematic symbols of Figure 4-2 that the emitter terminal, or lead, has an arrow, which is used to identify the type of transistor—NPN or PNP. If the arrow of the emitter is **N**ot **P**ointing i**N** as shown in Figure 4-2(a), it is an NPN transistor. If the arrow of the emitter is **P**ointing i**N P**ositively as shown in Figure 4-2(b), then it is a PNP transistor. A technician must be able to identify the type of transistor and each of the leads to test it and connect it in a circuit correctly.

Keep in mind that the schematic symbol doesn't indicate if a transistor is made of a germanium (Ge) or silicon (Si) base. Silicon transistors have almost

totally replaced germanium ones in electronic applications. Although germanium was the first element used in the manufacture of transistors, silicon transistors are easier and cheaper to manufacture, and they can handle more heat. Before we go into the operation of a transistor, we must first learn to identify and test a transistor.

4-2 IDENTIFYING TRANSISTORS USING A SEMICONDUCTOR REPLACEMENT GUIDE

Transistors come in a variety of sizes and shapes. If you encounter an electronic device with three terminals or leads, it would be a good guess that it's a transistor. However, not all transistors have three terminals, so we need to be sure. While many diode part numbers use the prefix 1N, transistor part numbers often have the prefix 2N. So, if an electronic device has three terminals and begins with the part number 2N, it's most likely a transistor. The only way to be absolutely sure is *Look it up to hook it up.*

Like diodes, every transistor has a manufacturer specification (spec) sheet that lists everything about the transistor. Figure 4-3 shows a spec sheet for a 2N3904 transistor. The spec sheet shows the transistor's diagram, applications, maximum rating, and thermal characteristics—anything you want to know about the transistor. Figure 4-3 shows only the first of seven sheets of information about the 2N3904 transistor.

Sometimes you can't access a spec sheet. Again, this is where a semiconductor replacement guide such as the *NTE* catalog comes in handy. The following example uses page numbers and information from the 14th edition of the *NTE* catalog, so your edition may have different page numbers. It's important to understand that the procedure is the same for any edition of the *NTE* catalog: start at the back of the book, go to the front of the book, and then go to the middle of the book. Back, front, middle.

For example, say you have a three-terminal electronic device with the part number 2N2219. You turn to the *back* of the *NTE* catalog and search for the part number 2N2219. (The back section of the catalog lists part numbers in alphanumeric order; that is, by letters and then numbers.) You will find that part number 2N2219 has an NTE replacement number of 123.

FAIRCHILD
SEMICONDUCTOR™

2N3904/MMBT3904/PZT3904

2N3904

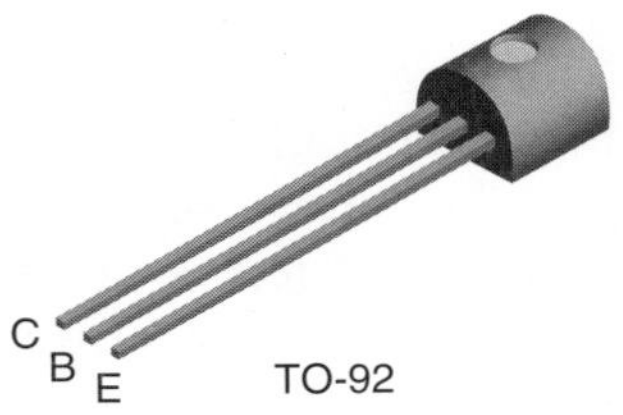
C
B
E
TO-92

MMBT3904

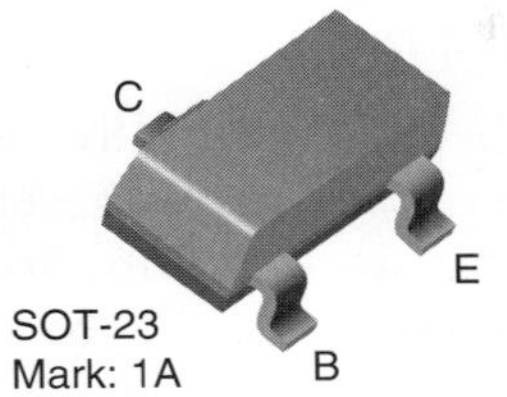
C
E
SOT-23
Mark: 1A
B

PZT3904

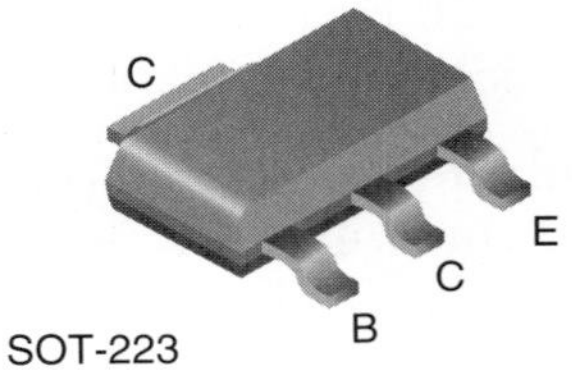
C
E
C
B
SOT-223

NPN General Purpose Amplifier

This device is designed as a general purpose amplifier and switch. The useful dynamic range extends to 100 mA as a switch and to 100 MHz as an amplifier.

Absolute Maximum Ratings* T_A = 25°C unless otherwise noted

Symbol	Parameter	Value	Units
V_{CEO}	Collector-Emitter Voltage	40	V
V_{CBO}	Collector-Base Voltage	60	V
V_{EBO}	Emitter-Base Voltage	6.0	V
I_C	Collector Current - Continuous	200	mA
T_J, T_{stg}	Operating and Storage Junction Temperature Range	−55 to +150	°C

*These ratings are limiting values above which the serviceability of any semiconductor device may be impaired.

NOTES:
1) These ratings are based on a maximum junction temperature of 150 degrees C.
2) These are steady state limits. The factory should be consulted on applications involving pulsed or low duty cycle operations.

Thermal Characteristics T_A = 25°C unless otherwise noted

Symbol	Characteristic	Max			Units
		2N3904	*MMBT3904	**PZT3904	
P_D	Total Device Dissipation Derate above 25°C	625 5.0	350 2.8	1,000 8.0	mW mW/°C
$R_{\theta JC}$	Thermal Resistance, Junction to Case	83.3			°C/W
$R_{\theta JA}$	Thermal Resistance, Junction to Ambient	200	357	125	°C/W

*Device mounted on FR-4 PCB 1.6" × 1.6" × 0.06."
**Device mounted on FR-4 PCB 36 mm × 18 mm × 1.5 mm; mounting pad for the collector lead min. 6 cm^2.

 2N3904/MMBT3904/PZT3904, Rev A

FIGURE 4-3 Specification sheet for a 2N3904 transistor (Courtesy of Fairchild Semiconductor®)

Now, you go to the *front* of the *NTE* catalog and look through the numerical index until you find the NTE type number 123, which is on page VII of the 14th edition. On the same line is additional information for the device: the page number, 1-18; the diagram number, 21a; and description, which includes *T* for transistor, and NPN.

If you turn to page 1-18 and look under NTE type number 123, you will see columns of information. This information includes the polarity (NPN) and material the transistor is made of (Si, or silicon); the description and application (Amp, Audio to VHF Freq., Sw); the case style (TO39); the diagram number, 21a; a variety of operating current, voltage, power, and frequency limits; and the typical forward current gain or h_{FE} (200), which we will need for testing the transistor. So, the electronic device with a part number 2N2219 is a silicon NPN transistor used as a switch or amplifier for audio and radio frequency applications. Audio frequencies are typically considered 20 Hz to 20,000 Hz; radio frequencies are frequencies above 20,000 Hz.

If you look at diagram 21a on page 1-66 of the *NTE* catalog, you will see drawings of the transistor and its dimensions plus a scheme for identifying the emitter, base, and collector terminals. The scheme for identifying the terminals or leads of many electronic devices is often called a **pin configuration** or pin layout or pin-out.

The drawing at the top shows the transistor from the side view, and the drawing at the bottom shows the transistor from the bottom view; that is, with the leads facing toward you. Notice the tab on the transistor. Manufacturers often make transistors with a tab close to the emitter terminal for quick identification. If we look at the table under the letter *a*, you will see that the terminal closest to the tab is 1 or the emitter, shown by the letter *E*. Terminal or lead 2 is the base (shown by the letter *B*), and terminal or lead 3 is the collector, shown by the letter *C*. Notice that the metal case of the transistor also has a letter *C*. This means that the collector terminal is connected internally to the metal case. Connecting the collector terminal to the case allows heat to escape easily from the device and keeps the transistor from overheating. You'll see that some transistors (especially power transistors) have only two terminals. For these, the collector *is* the metal case.

Of course, if your class does *not* have any *NTE Semiconductor Technical Guide and Cross Reference* catalogs, then you can use the on-line version of the *NTE* catalog. Logon to www.nteinc.com and type in part number 2N2219 in the Search box of the Component Cross Reference button to get the same information about the transistor that we found in the hard copy of the *NTE* catalog. Keep in mind that the main reason we use a semiconductor replacement guide is to identify the emitter, base, and collector terminals of a transistor so that we can test it or connect it properly in a circuit.

4-3 USING A DMM TO TEST TRANSISTORS

Once we've identified the emitter, base, and collector of a transistor, we can now test it to see if it's good. Looking again at Figure 4-2, you can see that

both the NPN and PNP transistors have *two* depletion regions created during the manufacturing process. (The diode has only one depletion region in the area between the P-type material and N-type material.) Because a BJT has two depletion regions, we'll have to perform additional steps to test it.

For testing purposes, an NPN transistor is really two PN junction diodes. One PN junction is present between the base lead and the emitter lead, and another PN junction is present between the base lead and the collector lead. Figure 4-4 shows a "phantom" diode (the blue triangle) between the base lead and the collector lead. This phantom diode is for learning purposes only.

Recall from checking a diode in Chapter 1 that the P-type or anode end of the diode is the positive end, and the N-type or cathode end of the diode is the negative end. We also learned that a force, or barrier voltage, exists between the P-type material and the N-type material, @ 0.3 V for a germanium diode and @ 0.6 V for a silicon diode. When you apply a positive voltage to the anode of a diode and a negative voltage to the cathode of a diode, the diode is forward biased (positive to positive and negative to negative), the barrier or depletion region breaks down, and current flows through the diode. On the other hand, when you apply a positive voltage to the cathode end of a diode and a negative voltage to the anode of a diode, the diode is reverse biased (positive to negative and negative to positive), the depletion region increases, and no current flows through the diode. These same concepts apply to the testing of a transistor.

One way to test a transistor is to use a digital multimeter (DMM) with a diode setting, which looks like the diode schematic symbol. In Chapter 1, we used the diode setting to test diodes, and since an

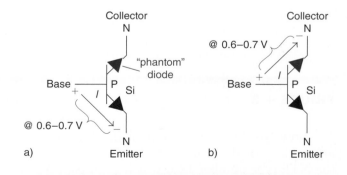

FIGURE 4-4 Forward biasing the depletion region in a silicon NPN transistor: a) Base-emitter forward biased and b) Base-collector forward biased (© Cengage Learning 2012)

NPN transistor is really two PN junction diodes, we can use it here. When you check a transistor on the diode setting, the numerical result is in volts, though the meter may *not* show "volts" or the letter *V*.

Testing a transistor on the diode setting requires six checks. First, set your meter's rotary selector switch to the diode setting. Then touch or clip the red (positive) lead of the meter to the emitter terminal of the transistor, and touch or clip the black (negative) lead of the meter to the base terminal of the transistor. This reverse biases the PN junction diode between the base and emitter, so the meter should display infinite, which may be represented by the infinity symbol (∞) or "OL" (overload) for a good transistor.

For the second check, touch or clip the red (positive) lead of the meter to the emitter of the transistor, and touch or clip the black (negative) lead of the meter to the collector of the transistor. Again, this reverse biases the PN junction diode between the collector and emitter, so the meter should display infinite, which may be represented by the infinity symbol (∞) or "OL" (overload) for a good transistor.

For the third check, touch or clip the red (positive) lead of the meter to the base of the transistor, and touch or clip the black (negative) lead of the meter to the emitter of the transistor. This forward biases the PN junction between the base and emitter, current flows through the depletion region, and the meter should display about 0.6 to 0.7 (volts) for a good silicon transistor as shown in Figure 4-4(a). If the transistor is made of germanium, the meter should display about 0.2 to 0.3 (volts).

For the fourth check, touch or clip the red (positive) lead of the meter to the base of the transistor, and touch or clip the black (negative) lead of the meter to the collector of the transistor. This forward biases the PN junction or phantom diode between the base and collector, current flows through the depletion region, and the meter should display about 0.6 to 0.7 (volts) for a good silicon transistor as shown in Figure 4-4(b). If the transistor is made of germanium, the meter should display about 0.2 to 0.3 (volts).

For the fifth check, touch or clip the red (positive) lead of the meter to the collector of the transistor, and touch or clip the black (negative) lead of the meter to the emitter of the transistor. This reverse biases the PN junction diode between the collector and emitter, so the meter should display infinite, which may be represented by the infinity symbol (∞) or "OL" (overload) for a good transistor.

For the sixth and final check, touch or clip the red (positive) lead of the meter to the collector of the transistor, and touch or clip the black (negative) lead of the meter to the base of the transistor. Again, this reverse biases the PN junction diode between the collector and base, so the meter should display infinite, which may be represented by the symbol (∞) or "OL" (overload) for a good transistor.

So, transistor testing using the diode setting and meter leads requires six checks. The results for testing an NPN transistor are summarized below:

1) Positive on emitter and negative on base—display should be infinite or OL.

2) Positive on emitter and negative on collector—display should be infinite or OL.

3) Positive on base and negative on emitter—display should be @ 0.6 to 0.7 (volts).

4) Positive on base and negative on collector—display should be @ 0.6 to 0.7 (volts).

5) Positive on collector and negative on emitter—display should be infinite or OL.

6) Positive on collector and negative on base—display should be infinite or OL.

Notice that there are two numerical readings (the result of forward biasing each junction) and four infinite readings for a good transistor. If the ratio is different—three and three or five and one or six and zero—then the transistor is defective and should be replaced. This six-check approach to testing a transistor has been used for over 50 years and is tried and true. Occasionally, however, a transistor can check good using the diode-setting/six-check method but then fail in a circuit under load conditions. The rule of thumb is: *When in doubt, swap it out.*

Many DMMs manufactured today have another method for testing BJTs that uses a setting/socket combination. Figure 4-5 shows a DMM with two settings for the label h_{FE}: NPN and PNP. Since BJTs are current-controlled devices, they will have a DC current gain, which will be explained in detail later in this chapter. For now, we just need to know that the DC current gain is a number and that it is also called h_{FE} or beta, often represented by the Greek letter β.

In Figure 4-5, if you look below the DMM selector switch to the left, you will also see the label h_{FE} and four little sockets labeled *E*, *B*, *C*, and *E*. The extra *E* on the far right is for a special transistor with two emitters; it's also used for transistors that

FIGURE 4-5 RSR Model 717 DMM on h$_{FE}$ setting, NPN (Courtesy of Electronix Express/RSR Electronics, Inc.)

have a pin configuration that follows the pattern *B*, *C*, and *E*. To check a transistor using this setting/socket method, set your meter's rotary selector switch to either the NPN or the PNP setting. Then insert the transistor leads in their correct sockets, matching the emitter lead to the emitter socket, the base lead to the base socket, and the collector lead to the collector socket. Then power on your DMM and observe the number display.

For example, we can use the 2N2219 transistor that we researched in the *NTE* catalog in the previous section. According to the *NTE* catalog, the 2N2219 is an NPN transistor that should have a typical forward current gain of 200. To test the transistor, set the DMM's rotary selector switch to NPN and insert the transistor leads into their correct sockets. Then power on the meter and observe the display. The meter display should be about 200 if the transistor is good. In some cases, you have to jiggle the transistor leads to get an accurate reading. The setting/socket method for transistor testing is really a ballpark approach because many transistors will read much lower than the manufacturer's specs but still work fine.

If an h$_{FE}$ reading is way off, check a similar transistor and compare the numbers.

Also, you can't use the h$_{FE}$ setting/socket method for high-power transistors because the transistor leads are too big to fit into the sockets. Personally, I prefer the older diode-setting/six-check method for testing transistors because I can check any size transistor using this approach. I tell my students to try both methods and use the one they feel more comfortable with.

To check a PNP transistor, you use the same methods. However, the meter display readings will be different because the PNP transistor is made of two parts of P-type material and one part N-type material. Testing a PNP transistor using the diode-setting/six-check method will show the following results for a good PNP transistor:

1) Positive on emitter and negative on base—display should be @ 0.6 to 0.7 (volts).

2) Positive on emitter and negative on collector—display should be infinite or OL.

3) Positive on base and negative on emitter—display should be infinite or OL.

4) Positive on base and negative on collector—display should be infinite or OL.

5) Positive on collector and negative on emitter—display should be infinite or OL.

6) Positive on collector and negative on base—display should be @ 0.6 to 0.7 (volts).

Notice that there are two numerical readings (the result of forward biasing each junction) and four infinite readings for a good transistor. If the ratio is different—three and three or five and one or six and zero—then the transistor is defective and should be replaced.

Of course, you can also check a PNP transistor using the h$_{FE}$ setting/socket method. To check a PNP transistor using this method, set your meter's rotary selector switch to the PNP setting. Then insert the transistor leads into their correct sockets, matching the emitter lead to the emitter socket, the base lead to the base socket, and the collector lead to the collector socket. Then power on your DMM and observe the number display.

Now that we know the methods for identifying and testing a transistor, it's time to put theory into practice.

Identifying and Testing Bipolar Junction Transistors (BJTs)

Materials, Equipment, and Parts:

- *NTE Semiconductor Technical Guide and Cross Reference* catalog

  or

  NTE QUICKCross™ software

  or

  Internet access, www. nteinc.com

- Digital Multimeter (DMM) with test leads.

- One BJT with case style TO92, such as 2N3904 or 2N3906.

- One BJT with case style TO39 (top hat), such as 2N2219 or 2N3053.

- One BJT with case style TO3 or TO66 (space-ship), such as 2N3055 or 2N6373.

Discussion Summary:

Bipolar Junction Transistors (BJTs) can go bad (become shorted or open). Technicians need to be able to identify the polarity of a transistor (NPN or PNP); the material it's made of (Si or silicon, Ge or germanium); and the emitter, base, and collector leads to test the transistor correctly. Two methods exist for testing a transistor: the diode-setting/six-check method and the h_{FE} setting/socket method. Some transistors can't be checked using the h_{FE} setting/socket method because their leads won't fit into the sockets.

(continues)

LAB ACTIVITY 4-1

(continued)

Procedure:

 Use the *NTE* catalog to research and record the following information for each of the three transistors at your workstation. If a transistor can't be checked using the h_{FE} setting/socket method in Step 4, place a dash (—) on the blank line. For the first transistor, Steps 1, 3, and 4 have been completed to serve as an example:

	1	2	3
Part #	2N3904	_____	_____
NTE #	123AP	_____	_____
Polarity	NPN	_____	_____
Material	Silicon	_____	_____
Description & Application	Audio amplifier, switch	_____	_____
Case Style	TO92	_____	_____
Diagram #	9a	_____	_____
h_{FE}	200	_____	_____

2 Go to the diagram page for each transistor and then draw the transistor outline (TO) for each transistor. Label the emitter, base, and collector terminals.

1   2   3

LAB ACTIVITY 4-1

3   Use your DMM's diode setting and meter leads to check each transistor. Record the results. For the first voltage reading, the positive lead (+) of the DMM is connected to the emitter terminal (E), and the negative lead (−) of the DMM is connected to the base terminal (B). Perform the remaining checks in this manner.

		1			**2**			**3**
+	−	Volt reading	+	−	Volt reading	+	−	Volt reading
E	B	OL	E	B	_____	E	B	_____
E	C	OL	E	C	_____	E	C	_____
B	E	0.697 V	B	E	_____	B	E	_____
B	C	0.689 V	B	C	_____	B	C	_____
C	E	OL	C	E	_____	C	E	_____
C	B	OL	C	B	_____	C	B	_____
Good?		Yes			_____			_____

4   Use your DMM's h_{FE} setting and socket to measure and record the h_{FE} (typical forward current gain) of each transistor.

	1	**2**	**3**
h_{FE} reading	167	_____	_____

5   How does the measured h_{FE} of Step 4 compare to the NTE-listed h_{FE} of Step 1 for each transistor?

6   Have the instructor check your results.

Using Solid State Challenges™ to Test Transistors

Using the PC at your workstation, open the Solid State Challenges™ to the lesson "Basic Transistor Testing." Read the introductory screen "Testing Transistors Using an Ohmmeter" and then click on the Begin tab. Read the directions in the red box, and then practice checking transistors until you complete all 12 exercises.

4-4 TRANSISTOR BIASING AND DC CURRENT GAIN (β)

Now that we understand how to identify and test a transistor, it's time to investigate exactly how a transistor works. The BJT is a current-controlled device. Since the base terminal behaves like a gate or valve, increasing the base current (I_B) increases the collector current (I_C). If there is no I_B, there is no I_C, and the transistor is off.

The I_B must be created by a DC voltage source, which is called bias. A BJT needs two DC voltages to operate correctly. When you bias a transistor, you are applying voltages to turn on the transistor. For proper operation, a BJT must have a forward bias on the base-to-emitter junction and a reverse bias on the base-to-collector junction.

Figure 4-6 shows a properly biased NPN transistor. This circuit is for learning purposes only.

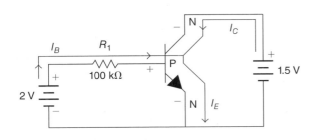

FIGURE 4-6 NPN transistor with base-to-emitter and base-to-collector biasing (© Cengage Learning 2012)

If you look carefully at Figure 4-6, you will see that the base-to-emitter junction is forward biased. The positive end of the 2 V DC source on the left is connected to the base terminal (made of P-type material), and the negative end of the 2 V DC source is connected to the emitter terminal, which is made of N-type material. Thus, we have positive to positive and negative to negative; that is, forward bias. On the right side of the transistor, the positive end of the 1.5 V DC source is connected to the collector terminal—made of N-type material—so the collector is reverse biased.

Recall that the emitter has a large number of excess electrons, the base has a very small number of excess holes, and the collector has a small number of excess electrons. How exactly does the circuit of Figure 4-6 operate?

The 2 V DC source on the left forward biases the base-to-emitter junction and breaks down the depletion region. Conventional, or hole, current flows through the 100 kΩ current-limiting resistor and into the base terminal. Hole current from the positive side of the 1.5 V DC source floods the collector terminal. The hole currents from the collector and the base combine and flow into the emitter. (Recall that the BJT's operation depends on the interaction of holes *and* electrons. Electron flow is opposite to hole current flow, so the excess electrons at the emitter join with electrons from the base and gather at the "collector" terminal.)

The amount of emitter current can be calculated using Equation 4-1.

EQUATION 4-1

Emitter current in BJT

$$I_E = I_C + I_B$$

I_E = current through emitter

I_C = current through collector

I_B = current through base

The actual currents in a BJT can range from μA to amps. For example, the 2N3904 transistor has a maximum I_C of 200 mA. Its NTE replacement, 123AP, has a maximum I_C of 600 mA. For a small transistor, the base current value could be in μA, and the emitter and collector current values could be in mA.

Since the base acts like a valve or gate to allow more electrons to flow into the collector, the I_B controls the I_C, and the BJT has a current gain. The **DC current gain** of a BJT is often called $\mathbf{h_{FE}}$, which we measured in the previous section. It is also called **beta**, often represented by the Greek letter β.

This DC current gain, or β, can be calculated using Equation 4-2.

EXAMPLE 1

Situation

If a transistor has an I_C of 50 mA and an I_B of 400 μA, what is the I_E?

Solution

First, we'll convert the I_B to get both numbers in the same unit: 400 μA = 0.4 mA

Then,

$$I_E = I_C + I_B = 50\,mA + 0.4\,mA$$

$$= 50.4\,mA$$

EQUATION 4-2

DC current gain transistor

$$\beta = \frac{I_C}{I_B}$$

I_C = collector current

I_B = base current

EXAMPLE 3

Situation

If a transistor has an I_C of 100 mA and an I_B of 800 μA, what is the β?

Solution

First, we'll convert the I_B to get both numbers in the same unit: 800 μA = 0.8 mA

Then,

$$\beta = \frac{I_C}{I_B} = \frac{100\,mA}{0.8\,mA} = 125$$

EXAMPLE 2

Situation

If a transistor has an I_C of 40.85 mA and an I_E of 41 mA, what is the I_B?

Solution

First, we start with the original equation, $I_E = I_C + I_B$.

We need to solve for I_B, so we rearrange the equation:

$$I_B = I_E - I_C = 41\,mA - 40.85\,mA$$

$$= 0.15\,mA \text{ or } 150\,μA$$

So the DC current gain of this transistor is 125, which means the I_C is 125 times greater than the I_B.

For Example 4, the DC current gain of this transistor is 85, which means the I_C is 85 times greater than the I_B. Keep in mind for both examples that β is a number without any unit.

So far, we've been saying that a small amount of base current can control a large amount of

EXAMPLE 4

Situation

If a transistor has an I_C of 425 mA and an I_B of 5 mA, what is the β?

Solution

Since both values are in mA, we don't have to convert.

$$\beta = \frac{I_C}{I_B} = \frac{425\,mA}{5\,mA} = 85$$

collector current. Let's use a real transistor and see just how much the I_B affects the I_C.

In Figure 4-7, we've added two ammeters to the circuit of Figure 4-6: one to measure the I_B and the other to measure the I_C. We've also added a variable DC voltage source on the left, set to 2 V. The transistor is labeled Q_1 because the letter T is used for transformers, which were invented before the transistor. The schematic designator for all transistors begins with the letter Q followed by a number that is usually in subscript form. For example, if you have three transistors in a circuit, they would be labeled Q_1, Q_2, and Q_3.

Link to Prior Learning

When a schematic symbol has an arrow through it, this means the electronic device's value is variable.

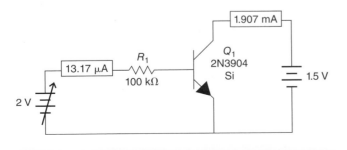

FIGURE 4-7 2N3904 NPN silicon transistor with base-to-emitter and base-to-collector biasing (© Cengage Learning 2012)

When the 2 V DC source on the left forward biases the base-to-emitter junction, the depletion region breaks down, and an I_B of 13.17 µA flows into the transistor. This causes an I_C of 1.907 mA. The resulting DC current gain or

$$\beta = \frac{I_C}{I_B} = \frac{1.907\,mA}{13.17\,\mu A} = 144.79$$

If the DC source is changed to 4 V, $I_B = 32.9$ µA, $I_C = 5.197$ mA, and $\beta = 157.96$

If the DC source is changed to 6 V, $I_B = 52.76$ µA, $I_C = 8.499$ mA, and $\beta = 161.08$

If the DC source is changed to 8 V, $I_B = 72.66$ µA, $I_C = 11.71$ mA, and $\beta = 161.16$

If the DC source is changed to 10 V, $I_B = 92.59$ µA, $I_C = 14.80$ mA, and $\beta = 159.84$

Let's consider the data. Overall, the bias, or DC voltage, varied from 2 V to 10 V, a change of 8 V. This bias caused the I_B to vary from 13.17 µA to 92.59 µA, a change of 79.42 µA. The I_B caused the collector current to vary from 1.907 mA to 14.80 mA, a change of 12.89 mA. When we consider the entire range, an I_B of 79.42 µA caused an I_C change of 12.89 mA. This proves that only a small change in I_B can cause a large change in I_C.

Also, look at the value of DC current gain, or β, for the entire range of DC voltage values. The β changed only by 16.37 (161.16 − 144.79). Thus, we can say that for a given transistor, the DC current gain, or β, is fairly constant. Of course, you can build the circuit of Figure 4-7 on Multisim®, Electronics Workbench®, or SPICE and then compare your measurements.

4-5 THE TRANSISTOR IN SWITCHING APPLICATIONS

We mentioned earlier that a transistor is used as a high-speed switch or as an amplifier. In Chapter 5, we'll cover the transistor as an amplifier. Now, we'll look at how the transistor can be used to rapidly switch a load on and off. Figure 4-8 shows a transistor in a switching application.

The circuit of Figure 4-8 has a function generator on the left that will supply an AC voltage

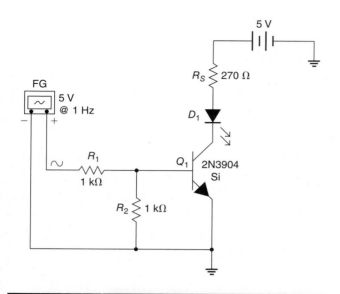

FIGURE 4-8   2N3904 NPN silicon transistor used as a high-speed switch (© Cengage Learning 2012)

When the AC voltage from the function generator is greater than @ 0.6 to 0.7 V across resistor R_2, the I_B causes I_C to flow, turning on the transistor. Current flows from the 5 V DC source on the right, through the 270 Ω resistor, through the LED, through the collector, through the base, through the emitter, and to ground. The LED has a positive voltage on its anode and a negative voltage on its cathode, so it lights.

When the AC voltage from the function generator drops below @ 0.6 V to 0.7 V across R_2, no I_B flows into the transistor, the transistor turns off, and the LED turns off. Thus, the alternating voltage from the function generator turns the transistor on and off. Initially, we have the function generator set to 1 Hz. If you increase the frequency of the function generator, then the LED will blink rapidly until you reach a frequency where your eyes see the LED as just one steady light. Let's build it live.

(sine wave) of 5 V @ 1 Hz to the base of the NPN silicon transistor. Resistor R_1 limits the I_B, and resistor R_2 supplies the bias across the base-to-emitter junction that will turn on the transistor. The LED connected to the collector terminal is the load. The 270 Ω resistor R_S limits the current through the LED.

The Transistor as a High-Speed Switch

Materials, Equipment, and Parts:

- *NTE* catalog or Internet access, www.nteinc. com (*Look it up to hook it up.*)

- PC w/Multisim®, Electronics Workbench®, or SPICE.

- DMM with test leads.

- 5 V DC voltage source.

- Function generator, 1 Hz to 60 Hz.

- Breadboard and connecting wires.

- 2N3904 or equivalent NPN transistor.

- Two-terminal LED (any color).

- 270 Ω fixed resistor.

- 1 kΩ fixed resistor (2).

Discussion Summary:

A transistor can be used as a high-speed switch. When an alternating voltage forward biases the base-to-emitter junction of a BJT, the I_B causes I_C to flow. If an LED is part of the collector circuit, then the LED will turn on if the incoming signal voltage goes above 0.6 V to 0.7 V. When the incoming signal voltage drops below 0.6 V, the transistor will turn off, turning off the LED.

Procedure:

SAFETY FIRST. Eye protection should always be worn when working with live voltages. Before powering on a live circuit, always check with your instructor.

1 Write the part number of the transistor at your workstation. Then use the *NTE* catalog or visit the website www.nteinc.com to find and record the NTE replacement number and the diagram number.

Part number _____

NTE number _____

Diagram # _____

2 Draw the transistor outline and label the emitter, base, and collector.

(continues)

LAB ACTIVITY 4-3

(continued)

3 Build the circuit of Figure 4-9 on your breadboard.

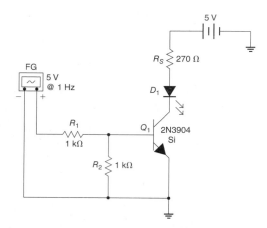

FIGURE 4-9   2N3904 NPN silicon transistor used as a high-speed switch (© Cengage Learning 2012)

4 Connect the positive lead of the function generator to the 1 kΩ base resistor.

5 Connect the positive lead of the 5 V DC voltage source to one end of the 270 Ω resistor.

6 Connect the black lead of the function generator, the black lead of the 5 V DC voltage source, the emitter of the transistor, and a wire from the bottom of resistor R_2 to one point (hole) on the breadboard, which is ground.

7 Set the function generator to the sine wave setting and to an amplitude of 5 V. You may have to use your DMM set on the AC volts setting to adjust the voltage to 5 V.

8 Set the function generator for a frequency of 1 Hz (or a little higher).

9 Have the instructor check your circuit.

10 Power on your 5 V DC voltage source and the function generator.

11 Does the LED turn on and off? Yes _____ No _____

12 Slowly increase the frequency of the function generator until the LED is a steady light. Record the frequency _____

13 Why does the LED become a steady light at the frequency of Step 12?

14 Have the instructor verify your results.

15 Build the circuit on Multisim®, Electronics Workbench®, or SPICE and then compare the results.

CHAPTER SUMMARY

The invention of the transistor in 1947 revolutionized electronics, spawning the solid-state or semiconductor industry. The transistor is smaller, more efficient, and more rugged than its vacuum tube counterpart, the triode.

The transistor is a three-terminal semiconductor device that is used as a high-speed switch or as an amplifier. Transistors come in three common transistor outlines or packages that you may encounter as an electronic technician.

The two major categories of transistors are Bipolar Junction Transistors (BJTs) and Field Effect Transistors (FETs). BJTs are current-controlled devices whose operation depends on the interaction of electrons and holes. They are constructed in one of two ways: NPN or PNP, and they have a base of Si or Ge. The silicon NPN transistor is most commonly used.

Transistor part numbers often have the prefix 2N, and each transistor has its own manufacturer specification (spec) sheet, which lists everything about the transistor.

Semiconductor replacement guides are used to identify the emitter, base, and collector terminals of a transistor so that it can be tested or connected properly in a circuit. Power transistors often have the collector terminal connected to the metal case to allow heat to escape easily and to keep the transistor from overheating.

For testing purposes, an NPN transistor is really two PN junction diodes. One way to test a transistor is to use test leads and a digital multimeter (DMM) set to the diode setting. You perform six checks: for a good transistor, four of these checks should display infinite, and two checks should display @ 0.6 V to 0.7 V for a good silicon transistor and @ 0.2 V to 0.3 V for a good germanium transistor, when the base-to-emitter and base-to-collector junctions are forward biased. Another way to test a transistor is to use the h_{FE} setting/socket method, which directly measures the DC current gain of the transistor.

The BJT is a current-controlled device. The base terminal behaves like a gate or valve, so increasing the base current (I_B) increases the collector current (I_C). The I_B must be created by a DC voltage source, which is called bias. For proper operation, a BJT must have a forward bias on the base-to-emitter junction and a reverse bias on the base-to-collector junction.

The transistor's collector current and base current combine and flow into the emitter. The emitter current is equal to the collector current plus the base current.

The DC current gain of a BJT is often called h_{FE}. It is also called beta, often represented by the Greek letter β, and is equal to the collector current divided by the base current. For a given transistor, β is fairly constant.

The transistor can be used as a high-speed switch. When an AC voltage applied to the base-emitter junction of a silicon transistor is greater than @ 0.6 V to 0.7 V, the I_B causes I_C to flow, turning on the transistor. A load connected to the collector circuit will also turn on. When the AC voltage from the function generator drops below @ 0.6 V to 0.7 V, no I_B flows into the transistor, the transistor turns off, and the load turns off.

CHAPTER EQUATIONS

Emitter current in BJT

(Equation 4-1)

$$I_E = I_C + I_B$$

I_E = current through emitter

I_C = current through collector

I_B = current through base

DC current gain transistor

(Equation 4-2)

$$\beta = \frac{I_C}{I_B}$$

I_C = collector current

I_B = base current

CHAPTER REVIEW QUESTIONS

Chapter 4-1

1. What vacuum tube device is similar to the transistor?

2. Name the two general uses or applications of a transistor.

3. What are the two major categories of transistors?

4. Is the BJT a current-controlled device or a voltage-controlled device?

5. Draw and label the schematic symbols for an NPN and a PNP transistor.

6. Why are most transistors made from a silicon base rather than a germanium base?

Chapter 4-2

7. What part number prefix is commonly used to identify diodes? What prefix is commonly used to identify transistors?

8. Is the 2N3906 an NPN or a PNP transistor?

9. What is the maximum collector current for transistor 2N3055?

10. Draw and label the transistor outline for part number 2N6373.

Chapter 4-3

11. How many checks are required to test a BJT using a DMM's diode setting and test leads?

12. Of these checks in Question 11, how many should show a reading of 0.6 V to 0.7 V for a good silicon transistor?

13. What are the two other names for the DC current gain of a BJT?

14. Why can't some transistors be checked using the h_{FE}/socket method?

Chapter 4-4

15. What type of bias must be applied to the base-to-emitter junction of a BJT, forward or reverse bias?

16. What type of bias must be applied to the base-to-collector junction of a BJT, forward or reverse bias?

17. $I_C = 3.71$ mA
    $I_B = 455$ µA
    $I_E = ?$

18. $I_E = 14$ mA
    $I_B = 800$ µA
    $I_C = ?$

19. What is the DC current gain for Figure 4-10?

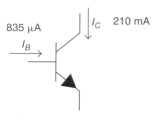

835 µA
I_B
I_C 210 mA

FIGURE 4-10 NPN silicon transistor with given values of I_C and I_B (© Cengage Learning 2012)

20. $I_C = 130$ mA
    $I_B = 690.2$ µA
    $\beta = ?$

21. $I_C = 601$ mA
    $\beta = 180$
    $I_B = ?$

22. What letter is used as the schematic designator for all transistors?

Chapter 4-5

23. If the I_B of a BJT is 0 A, what is the I_C?

24. What four types of signals are typically available at the output of a function generator?

25. How much base-to-emitter voltage is needed to turn on a germanium transistor?

5

Amplifier Configurations

OBJECTIVES   *Upon completion of this chapter, you should be able to:*

- Define amplification.
- Draw and label the generic schematic symbol for an amplifier.
- Explain the purpose of impedance matching.
- Calculate the power gain, current gain, and voltage gain of an amplifier.
- Explain how voltage-divider bias provides DC voltages to turn on a transistor amplifier.
- Distinguish among the DC voltages used to bias a transistor amplifier.
- Identify a transistor amplifier configuration from a schematic.
- List the distinctive features of the common-emitter (CE), common-collector (CC), and common-base (CB) amplifier configurations.

- Explain the function of a bypass capacitor in a CE amplifier.
- Calculate the DC voltage drops and static voltage gain of a CE amplifier.
- Build a CE amplifier circuit to calculate and measure its dynamic voltage gain.
- Build a CC amplifier circuit to calculate and measure its dynamic voltage gain.
- Calculate the decibel (dB) voltage and power gains of an amplifier.

MATERIALS, EQUIPMENT, AND PARTS

Materials, equipment, and parts needed for the lab experiments in this chapter are listed below:

- *NTE* catalog or Internet access, www.nteinc.com
- PC w/Multisim®, Electronics Workbench®, or SPICE.
- PC w/Solid State Challenges™.
- DMM with test leads.
- Dual trace oscilloscope w/BNC-to-alligator leads.

- 12 V DC voltage source.
- Function generator, 1 Hz to 10 kHz.
- Breadboard and connecting wires.
- 2N3904 or equivalent NPN transistor.
- 0.1 µF capacitor, 1 µF electrolytic capacitor (2), 10 µF

electrolytic capacitor, and 100 µF electrolytic capacitor.
- 470 Ω fixed resistor, 1 kΩ fixed resistor, 1.8 kΩ fixed resistor (2), 10 kΩ fixed resistor, 15 kΩ fixed resistor, and 20 kΩ fixed resistor.

GLOSSARY OF TERMS

Amplifier An electronic device or circuit that takes a small signal (usually AC) and makes it bigger; it can be a power, current, or voltage amplifier

Gain In an amplifier circuit, the result of the output signal divided by the input signal; that is, how many times bigger the output signal is than the input signal; it can be a power, current, or voltage gain

Voltage-divider bias An amplifier biasing scheme where the main DC power supply voltage is divided among various resistors to turn on a transistor so that it's ready for amplification

V_{CC} The main DC supply voltage for a BJT amplifier, which is a positive voltage applied across the entire transistor with respect to ground

Amplifier configuration The type of transistor scheme determined by the relationship between the AC input signal and the AC output signal

Common-emitter amplifier The most popular BJT amplifier configuration, which has the input signal applied between the base terminal and emitter terminal and

the output signal appearing between the collector terminal and the emitter terminal

Common-collector amplifier The BJT amplifier configuration that has the input signal applied between the base terminal and the collector terminal and the output signal appearing between the emitter terminal and the collector terminal; it is also called an emitter follower

Common-base amplifier The BJT amplifier configuration that has the input signal applied between the emitter terminal and the base terminal and the output signal appearing between the collector terminal and the base terminal

Bypass capacitor A capacitor connected across the emitter resistor in a common-emitter configuration that diverts to ground any AC signal appearing on the emitter terminal; it maintains the voltage gain of a common-emitter amplifier

Static voltage gain The calculated gain of an amplifier when no AC signal is applied

Dynamic voltage gain The gain of an amplifier with an AC signal applied; the result of the output

voltage divided by the input voltage

Emitter follower BJT amplifier configuration where the output voltage tracks or "follows" the amplitude and phase of the input voltage; it is also known as a common-collector amplifier

Decoupling capacitor A capacitor connected in series with the DC power supply voltage (V_{CC}) of a common-collector amplifier; it diverts from V_{CC} any unwanted AC on the collector terminal

Darlington amplifier or Darlington pair A semiconductor amplifier that uses two common-collector transistors in one housing to provide a large current gain

Emitter bias An amplifier biasing scheme used less often than voltage-divider bias because it requires two power supplies, one positive and one negative

Decibel (dB) A unit of measurement that uses the logarithm function and the ratio of two units to calculate an amplifier's gain or loss; it expresses gain in a manageable quantity

5-1 CHARACTERISTICS OF AMPLIFIERS

In Chapter 4, we learned how a transistor operates and how to connect it as a high-speed switch. A transistor is also used as an amplifier. An **amplifier** is an electronic device that takes a small signal (usually AC) and makes it bigger. The generic schematic symbol for an amplifier is a triangle with one point facing to the right as shown in Figure 5-1(a).

The input signal enters the left side of the triangle, and the output signal exits the point of the triangle. The three most important characteristics of an amplifier are input impedance (Z_{IN}), output impedance (Z_{OUT})—both shown in Figure 5-1(b)—and gain—shown in Figure 5-1(c).

A Citizens' Band (CB) radio transceiver is a good example of an amplifier having input and output impedances. The incoming radio wave on the antenna "sees" the CB radio as an AC resistance, or input impedance. On the other end of the amplifier, external speakers connected to the CB radio "see" the CB radio as an AC resistance, or output impedance. For maximum transfer of power, the external devices connected to a CB radio must match the input and output impedances of the CB radio.

For example, most CB antennas have an impedance of 50 Ω or 75 Ω. If the Z_{IN} of the CB radio is 50 Ω, then you should use an antenna with an impedance of 50 Ω to get the best transmission and reception and to waste less power. Also, most external speakers have input impedances of 4 Ω, 8 Ω, or 16 Ω. If the Z_{OUT} of your CB radio is 8 Ω, then you should connect 8 Ω speakers to the CB radio output for maximum transfer of power.

A technician is also concerned with the gain of an amplifier because one can calculate, measure, and change the gain. The **gain** of an amplifier is simply the output signal divided by the input signal; that is, how many times bigger the output signal is than the input signal.

$$\text{Amplifier gain} = \frac{\text{OUT}}{\text{IN}}$$

To be more specific, amplifiers can have three types of gain: power gain (A_P), current gain (A_I), and voltage gain (A_V). The letter A is used to represent amplification or gain, which is a little confusing since A is also used to represent ampere, the unit for current. These three gains are AC gains, not to be confused with the DC current gain of a transistor, which is called β or h_{FE}. Let's look at these gains one at a time.

EQUATION 5-1

Amplifier power gain

$$A_P = \frac{P_{OUT}}{P_{IN}}$$

EXAMPLE 1

Situation

An amplifier has an output power of 2.5 W and an input power of 40 mW. What is the power gain?

Solution

$$A_P = \frac{P_{OUT}}{P_{IN}} = \frac{2.5\,W}{40\,mW} = 62.5$$

For Example 5-1, the output power is 62.5 times greater than the input power. Notice that gain is just a number; gain has no unit. Like the DC current gain (β) of a transistor, gain is just a multiplier.

a) Input ———▷——— Output

b) Source (antenna) ▭—▷— Z_{IN} —AMP— Z_{OUT} —◁—▭ Load (speaker)

c) Input signal ∼ —AMP▷— ∼ Output signal

Gain = $\dfrac{\text{Out}}{\text{In}}$

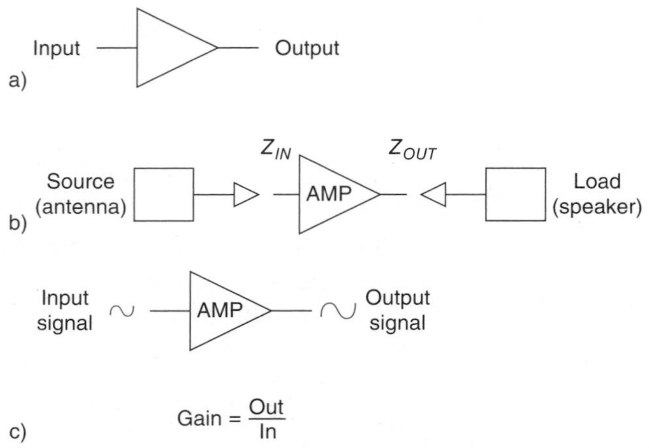

FIGURE 5-1 Amplifier symbol and characteristics: a) Schematic symbol, b) Z_{IN} and Z_{OUT}, and c) Gain (© Cengage Learning 2012)

EQUATION 5-2

Amplifier current gain

$$A_I = \frac{I_{OUT}}{I_{IN}}$$

EXAMPLE 2

Situation

An amplifier has an output current of 35 mA and an input current of 175 µA. What is the current gain?

Solution

$$A_I = \frac{I_{OUT}}{I_{IN}} = \frac{35\,mA}{175\,\mu A} = 200$$

For Example 5-2, the output current is 200 times greater than the input current. Again, gain is just a number.

EQUATION 5-3

Amplifier voltage gain

$$A_V = \frac{V_{OUT}}{V_{IN}}$$

EXAMPLE 3

Situation

An amplifier has an output voltage of 6 V and an input voltage of 50 mV. What is the voltage gain?

Solution

$$A_V = \frac{V_{OUT}}{V_{IN}} = \frac{6\,V}{50\,mV} = 120$$

For Example 5-3, the output voltage is 120 times greater than the input voltage. Again, gain is just a number.

Every amplifier has at least one of these gains, some have two types of gains, and one has all three. Many transistor amplifiers are designed specifically for a certain type of gain. Before any transistor can provide amplification, however, it needs bias (a DC voltage). We'll cover bias next.

5-2 AMPLIFIER BIASING

A transistor amplifier usually amplifies an AC signal, but first it needs a DC voltage (bias) to supply its operating power. The most common and preferred method used to bias transistor amplifiers is called voltage-divider bias. In **voltage-divider bias**, the main DC power supply voltage is divided among various resistors to turn on a transistor so that it's ready for amplification. Figure 5-2 shows voltage-divider bias and the distribution of DC voltages in a typical BJT amplifier.

The main DC supply voltage to a transistor amplifier is called the **V$_{CC}$**, which is a positive voltage applied across the entire transistor with respect to ground. For Figure 5-2, the V$_{CC}$ is 18 V.

As far as voltage distribution goes, it's as though you can divide a transistor in half—a left side and a right side. The blue dotted line in Figure 5-2 shows this division.

On the left side, the V$_{CC}$ is divided between resistors R$_1$ and R$_2$. Resistor R$_1$'s main job is to drop a large portion of the V$_{CC}$, and you can see

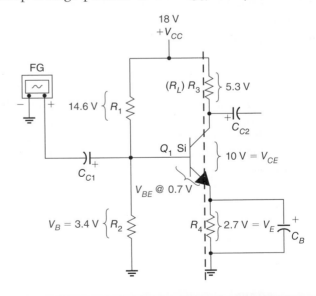

FIGURE 5-2 Voltage-divider bias and distribution of DC voltages in a BJT amplifier (© Cengage Learning 2012)

that 14.6 V appears across R_1. The remainder of the V_{CC} (3.4 V) is dropped across resistor R_2. Resistor R_2 is often called the biasing resistor because it applies the forward voltage across the base-to-emitter junction to turn on the transistor: @ 0.2 V to 0.3 V if the transistor is of germanium (Ge) and @ 0.6 V to 0.7 V if the transistor is of silicon (Si). We'll use 0.3 V and 0.7 V for our calculations. The transistor in Figure 5-2 is made of silicon.

On the right side of the transistor in Figure 5-2, the V_{CC} is divided among two resistors and the transistor itself. Resistor R_3 is often called the load resistor R_L, and it has 5.3 V dropped across it. When the transistor is on, it is also behaving like a resistor, so it drops a voltage, which is called V_{CE}. In Figure 5-2, the transistor has 10 V appearing across it. Finally, resistor R_4, which is connected to the emitter, drops 2.7 V. This voltage across the emitter resistor is often called V_E.

All the DC voltages on the left side of the transistor add up to the total voltage, or V_{CC} ($V_{R1} + V_{R2} = V_{CC}$). Also, all the DC voltages on the right side of the transistor add up to the total voltage, or V_{CC} ($V_{R3} + V_{CE} + V_E = V_{CC}$). Sounds familiar?

Link to Prior Learning

Kirchhoff's Voltage Law

In a series DC circuit, the total voltage V_T equals the sum of the voltage drops.

$$V_T = V_{R1} + V_{R2} + V_{R3} + V_R \cdots$$

Yes, the biasing of a transistor amplifier is another application of Kirchhoff's Voltage Law because bias involves DC voltages and resistors that are connected in series. Remember, a transistor is powered by DC.

The transistor amplifier DC voltages are summarized below:

V_{CC} Positive collector supply voltage; supplies voltage for entire transistor.

V_B Voltage between the base terminal and ground.

V_{BE} Voltage between the base and emitter terminals; forward biases the transistor.

V_{RL} Voltage drop across the load resistor.

V_{CE} Voltage between the collector and emitter (voltage drop across entire transistor).

V_E Voltage between the emitter and ground (voltage drop across emitter resistor).

V_C Voltage between the collector terminal and ground.

One last point about biasing. Notice in the DC voltages above that when one subscript letter is used (such as V_B, V_E, and V_C), the voltage is measured from the transistor terminal to ground. When two *different* subscript letters are used (V_{BE}, V_{RL}, and V_{CE}), the voltage is measured across two terminals or two points. For example, V_{CE} is measured with the positive lead of your voltmeter connected to the collector terminal and with the negative lead of your voltmeter connected to the emitter terminal.

V_{CC} is an odd case. The main DC supply voltage for a transistor is identified by two *identical* subscript letters (V_{CC}), but voltage is measured from the V_{CC} point to ground.

5-3 COMMON-EMITTER AMPLIFIER

The three transistor amplifier configurations are the common-emitter, the common-collector, and the common-base. A transistor **amplifier configuration** is determined by the relationship between the AC input signal and the AC output signal. Remember, a bias or DC voltage supplies the operating power for a transistor, but a transistor amplifies an AC signal.

For example, if the input signal is applied between the base terminal and emitter terminal and if the output signal appears between the collector terminal and the emitter terminal, then the configuration is called a **common-emitter amplifier** since the emitter is common to both the input signal and the output signal. It will have the features shown in the top row of Table 5-1.

Using the *same* transistor, if the input signal is applied between the base terminal and the collector terminal and if the output signal appears between the emitter terminal and the collector terminal, then the configuration is called a **common-collector amplifier** (collector is common to both the input signal and the output signal). It will have the features shown in the middle row of Table 5-1.

TABLE 5-1 Amplifier configurations (relationship between input and output signals) (© Cengage Learning 2012)

CONFIGURATION	VISUAL / PHYSICAL ASPECT	INPUT IMPEDANCE	OUTPUT IMPEDANCE	VOLTAGE GAIN	CURRENT GAIN	POWER GAIN	SPECIAL FEATURES
CE (Common-Emitter)	Input – Base Output – Collector	Midrange	Midrange	Midrange	Midrange	High	180° degree phase shift; is most common
CC (Common-Collector)	Input – Base Output – Emitter	High	Low	≤1	Midrange	Midrange	Emitter follower
CB (Common-Base)	Input – Emitter Output – Collector	Low	High	Midrange	≤1	Midrange	Input for high frequency receivers

Notes: 1) Midrange impedance typically means between 1 kΩ and 10 kΩ, so a low impedance would be below 1 kΩ, and a high impedance would be greater than 10 kΩ.

2) Midrange gain typically means between 100 and 1000, so a low gain would be below 100, and a high gain would be greater than 1000.

Again, using the *same* transistor, if the input signal is applied between the emitter terminal and the base terminal and if the output signal appears between the collector terminal and the base terminal, then the configuration is called a **common-base amplifier** (base is common to both the input signal and the output signal). It will have the features shown in the bottom row of Table 5-1.

Confusing? Let's look at another way to identify a transistor's configuration. In the common-emitter configuration, the AC signal enters at the base and exits from the collector. The AC signal does not appear at the emitter terminal, so it's a common-*emitter* amplifier. For the common-collector configuration, the AC signal enters at the base and exits from the emitter. The AC signal does not appear at the collector terminal, so it's a common-*collector* amplifier. For the common-base configuration, the AC signal enters at the emitter and exits from the collector. The AC signal does not appear at the base terminal, so it's a common-*base* amplifier. In all three cases, the terminal of the transistor that doesn't touch the input signal or output signal determines the configuration *name*.

The common-emitter (CE) is the most widely used amplifier configuration. Figure 5-3 shows a common-emitter amplifier with resistor and capacitor values.

Notice that the AC input signal is applied to the base terminal and the AC output signal

appears at the collector terminal. The AC signal at the output terminal is "flipped over" when compared to the AC input signal; this is called a 180° phase shift. In other words, the output signal is reaching its positive peak when the input signal is reaching its negative peak and vice versa. As shown in Table 5-1, this is one of the features of the common-emitter amplifier. Looking at Table 5-1, we also see that the common-emitter has midrange values for most of its operating characteristics. This also makes it desirable in amplifiers because the common-emitter amplifier can be connected to other amplifier configurations without having to make major adjustments. Many electronic systems such as a radio receiver have multiple amplifier stages.

Since the common-emitter is the most commonly used amplifier configuration, we'll analyze and determine its voltage gain without an AC signal applied, and then we'll build a common-emitter amplifier on the breadboard and determine its voltage gain with an AC signal applied.

In Figure 5-3, the V_{CC} supplies 16 V DC to the entire transistor with respect to ground. C_{C1} and C_{C2} are coupling capacitors that are used to bring the AC signal in and out of the transistor amplifier. An amplifier doesn't make an AC signal bigger by magic. The AC signal "steals" power from the V_{CC} to get bigger, but then the AC signal doesn't want to have anything to do with DC once it has gotten its gain. Consequently, C_{C2} keeps any DC within the amplifier stage.

Note, too, that the emitter resistor R_4 has a capacitor C_B in parallel with it. C_B stands for **bypass capacitor**. For maximum gain in a common-emitter amplifier, the AC signal should appear only at the collector terminal. If any of the AC signal leaks through the emitter terminal, it will lower the voltage gain of the amplifier. Thus, if any AC signal appears on the emitter terminal, the job of capacitor C_B is to bypass the emitter terminal and dump this AC to ground where it will reenter the base terminal through resistor R_2. This bypass capacitor is needed to maintain the voltage gain of a common-emitter amplifier.

Let's do an analysis to determine the voltage gain of the common-emitter amplifier. The **static voltage gain** of an amplifier is calculated when no AC signal is applied. Several steps are needed to calculate the static voltage gain of the common-emitter amplifier in Figure 5-3. Notice the transistor

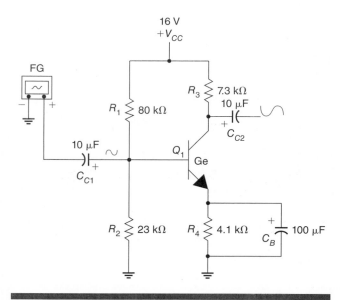

FIGURE 5-3 Common-emitter amplifier with Ge transistor (© Cengage Learning 2012)

is made of Ge (germanium). We'll start on the left side of the transistor.

1) Resistors R_1 and R_2 are in series with the DC supply voltage V_{CC}, so we can use the voltage divider equation from DC class to find V_{R1}.

$$V_{R1} = \frac{(R_1)(V_{CC})}{R_1 + R_2} = \frac{(80\,k\Omega)(16\,V)}{80\,k\Omega + 23\,k\Omega} = 12.42\,V$$

2) The same equation applies to V_{R2}.

$$V_{R2} = \frac{(R_2)(V_{CC})}{R_1 + R_2} = \frac{(23\,k\Omega)(16\,V)}{80\,k\Omega + 23\,k\Omega} = 3.57\,V$$

3) Recall from Figure 5-2 that the voltages drops on the left side of the transistor add up to the total voltage, or V_{CC}.

$$V_{R1} + V_{R2} = V_{CC}$$

V_{R1} and V_{R2} should add up to the total voltage, or V_{CC}, so we do a check.

Check: 12.42 V + 3.57 V = 15.99 V. That's close enough, so we're good so far.

Now, we go to the right side of the transistor.

4) The voltage on the emitter resistor should be equal to the voltage across the biasing resistor R_2 minus the 0.3 V dropped across the base-to-emitter junction (since it's Ge) that turns on the transistor.

$$V_E = V_{R2} - V_{BE} = 3.57\,V - 0.3\,V = 3.27\,V$$

5) Since we have the voltage across the emitter resistor R_4 and we know its resistance value, we can figure the current through the emitter, I_E.

$$I_E = \frac{V_E}{R_4} = \frac{3.27\,V}{4.1\,k\Omega} = 0.000797\,A = 0.797\,mA$$

Now, we take a little leap of faith. Remember, for a transistor the I_E and I_C are about the same, usually in mA. Thus, we can say the I_E is about equal to the I_C ($I_E \approx I_C$), so now we know the I_C value.

$$I_C = 0.797\,mA$$

6) Since we know the I_C value is 0.797 mA and the collector resistor value is 7.3 kΩ, we can use

Ohm's Law to calculate the voltage across the collector resistor R_3, also known as the load resistor.

$$V_{R3} = (I_C)(R_3) = (0.797\,mA)(7.3\,k\Omega) = 5.82\,V$$

7) Remember from Figure 5-2 that all the voltage drops on the right side of the transistor also add up to the total voltage, or V_{CC}.

$$V_{R3} + V_{CE} + V_E = V_{CC}$$

Well, we can rewrite this equation to find V_{CE}, the voltage dropped across the transistor when it's on.

$$V_{CE} = V_{CC} - V_{R3} - V_E = 16\,V - 5.82\,V - 3.27\,V$$
$$= 6.91\,V$$

So far, we've calculated all the DC voltages that we need using concepts and equations learned in DC class. Now, to figure the static gain of a transistor amplifier, we need to know the AC emitter resistance, or r'e (pronounced "are prime eee"). The AC emitter resistance is an engineering estimate of the emitter resistance when an AC signal is applied. We use this estimate because we can't measure any resistance with power applied and because we're calculating the static gain of the transistor amplifier; that is, without an AC signal at its input.

EQUATION 5-4

Amplifier AC emitter resistance

$$r'e = \frac{25\,mV}{I_E}$$

8) $$r'e = \frac{25\,mV}{I_E} = \frac{25\,mV}{0.797\,mA} = 31.36\,\Omega$$

Now, we need one last equation to calculate the gain.

EQUATION 5-5

Amplifier static voltage gain

$$A_V = \frac{R_3}{r'e}$$

9) $$A_V = \frac{R_3}{r'e} = \frac{7.3\,k\Omega}{31.36\,\Omega} = 232.78$$

EXAMPLE 4

Situation

Calculate the DC voltage drops and the static voltage gain for the transistor amplifier in Figure 5-4. Notice the transistor is made of Si (silicon). Again, we'll start on the left side of the transistor.

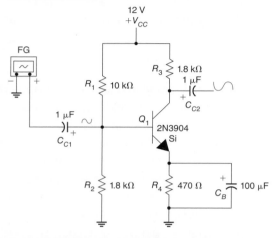

FIGURE 5-4   Common-emitter amplifier with Si transistor (© Cengage Learning 2012)

Solution

1) $V_{R1} = \dfrac{(R_1)(V_{CC})}{R_1 + R_2} = \dfrac{(10\,k\Omega)(12\,V)}{10\,k\Omega + 1.8\,k\Omega}$

$$= 10.16\,V$$

2) $V_{R2} = \dfrac{(R_2)(V_{CC})}{R_1 + R_2} = \dfrac{(1.8\,k\Omega)(12\,V)}{10\,k\Omega + 1.8\,k\Omega}$

$$= 1.83\,V$$

3) *Check*: $V_{R1} + V_{R2} = V_{CC}$

10.16 V + 1.83 V = 11.99 V. Close enough. Now, we'll go to the right side of the transistor.

4) The voltage on the emitter resistor should be equal to the voltage across the biasing resistor R_2 minus the 0.7 V dropped across the base-to-emitter junction (since it's Si) that turns on the transistor.

$$V_E = V_{R2} - V_{BE} = 1.83\,V - 0.7\,V$$

$$= 1.13\,V$$

5) $I_E = \dfrac{V_E}{R_4} = \dfrac{1.13\,V}{470\,\Omega} = 0.00240\,A$

$$= 2.40\,mA$$

Since $I_E \approx I_C$, $I_C = 2.40\,mA$

6) $V_{R3} = (I_C)(R_3) = (2.40\,mA)(1.8\,k\Omega)$

$$= 4.32\,V$$

7) $V_{R3} + V_{CE} + V_E = V_{CC}$

$V_{CE} = V_{CC} - V_{R3} - V_E$

$$= 12\,V - 4.32\,V - 1.13\,V = 6.55\,V$$

8) $r'e = \dfrac{25\,mV}{I_E} = \dfrac{25\,mV}{2.40\,mA} = 10.41\,\Omega$

9) $A_V = \dfrac{R_3}{r'e} = \dfrac{1.8\,k\Omega}{10.41\,\Omega} = 172.91$

We made it! The static voltage gain of the transistor amplifier of Figure 5-3 is 232.78, which means the AC output signal should be 232.78 times greater than the AC input signal. This is a typical midrange A_V (@ 100 to 1000). Let's do another amplifier analysis before we build one live.

The static voltage gain of the transistor amplifier of Figure 5-4 is 172.91, which means the AC output signal should be 172.91 times greater than the AC input signal.

An amplifier's static voltage gain is a ballpark calculation. The **dynamic voltage gain** of an amplifier is the actual measured gain. When a transistor is powered by DC and is amplifying an AC signal, then the dynamic voltage gain can be determined. For the dynamic voltage gain, you measure the V_{IN}, the V_{OUT}, and then use Equation 5-3 to determine the A_V:

$$A_V = \frac{V_{OUT}}{V_{IN}}$$

In an upcoming lab activity, we'll calculate the static gain for a common-emitter amplifier. Then we'll build the amplifier circuit and supply it with DC power. Finally, we'll apply an AC signal to the amplifier and determine the dynamic voltage gain. Keep in mind the following tip during the lab activity.

Troubleshooting Tip

The first step in troubleshooting is to do a visual inspection. If an amplifier doesn't show the correct V_{OUT}, any V_{OUT}, or a gain, first check your V_{CC} and ground. Then check the bypass capacitor. Next, check the resistors that provide the bias.

If you want more practice in calculating amplifier values, you can do the exercises in Lab Activity 5-1, or you can move on to Lab Activity 5-2 where you will build a common-emitter amplifier live, measure the V_{IN} and V_{OUT} to determine the A_V, and then compare the results to the static calculations.

LAB ACTIVITY 5-1

Using Solid State Challenges™ to Analyze Common-Emitter Amplifiers

Using the PC at your workstation, open the Solid State Challenges™ to the lesson "Common Emitter Amplifiers Three (Voltage-Divider Bias with Emitter Bypass Capacitor)." Read the introductory screen and then click on the Begin tab. Read the directions in the red box, and then perform all the calculations to determine the static gain of a common-emitter amplifier.

LAB ACTIVITY 5-2

Common-Emitter Amplifier

Materials, Equipment, and Parts:

- *NTE* catalog or Internet access, www.nteinc. com (*Look it up to hook it up.*)

- PC w/Multisim®, Electronics Work-bench®, or SPICE.

- DMM with test leads.

- Dual trace oscilloscope w/BNC-to-alligator leads.

- 12 V DC voltage source.

- Function generator, 1 Hz to 10 kHz.

- Breadboard and connecting wires.

- 2N3904 or equivalent NPN transistor.

- 1 µF electrolytic capacitor (2) and 100 µF electrolytic capacitor.

- 470 Ω fixed resistor, 1.8 kΩ fixed resistor (2), and 10 kΩ fixed resistor.

Discussion Summary:

A transistor can be used as a common-emitter (CE) amplifier. The CE has an AC input signal applied to the base terminal, and the AC output signal appears at the collector terminal. One feature of the CE is a 180° phase shift of the AC signal; the AC signal at the output terminal is flipped over when compared to the AC input signal. That is, the output signal is reaching its positive peak when the input signal is reaching its negative peak and vice versa. Another feature of the CE is a midrange voltage gain, or A_V, of 100 to 200. In this lab activity, we'll calculate the static gain and then measure the dynamic gain of the CE amplifier.

Procedure:

SAFETY FIRST. Eye protection should always be worn when working with live voltages. Before powering on a live circuit, always check with your instructor.

1 Write the part number of the transistor at your workstation. Then use the *NTE* catalog or visit the website www.nteinc.com to find and record the NTE replacement number and the diagram number.

Part number _____

NTE number _____

Diagram # _____

2 Draw the transistor outline and label the emitter, base, and collector.

(continues)

(continued)

3 Build the circuit of Figure 5-5 on your breadboard.

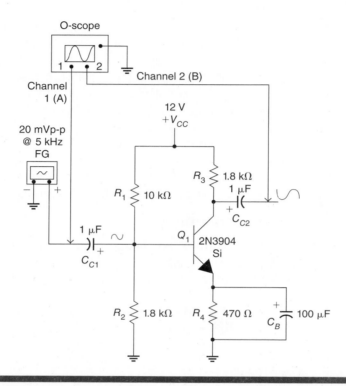

FIGURE 5-5 Common-emitter amplifier with Si transistor (© Cengage Learning 2012)

4 Connect the positive lead of the 12 V DC voltage source to the top end of resistors R_1 and R_3.

5 Connect the positive lead of the function generator to the negative side of capacitor C_{C1}.

6 Connect the positive lead of the oscilloscope's Channel 1 to the negative side of capacitor C_{C1}. Connect the positive lead of the o-scope's Channel 2 to the negative side of capacitor C_{C2}.

LAB ACTIVITY 5-2

7 Connect the black lead of the function generator, the black lead of the 12 V DC voltage source, the black leads of the o-scope, a wire from the bottom of resistor R_2, and a wire from the bottom of resistor R_4 to one point (hole) on the breadboard, which is ground. Figure 5-6 shows the actual breadboard circuit.

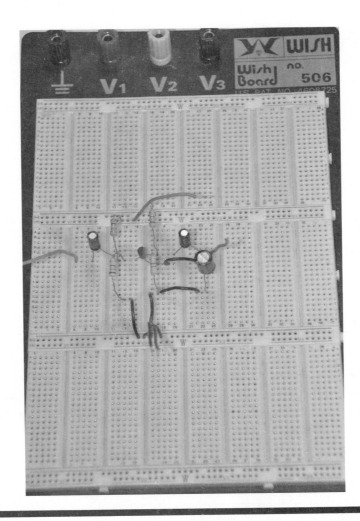

FIGURE 5-6   Common-emitter amplifier breadboard circuit (Courtesy of Tracy Grace Leleux)

8 Set the function generator to the sine wave setting and to an amplitude of 20 mV peak-to-peak. You may have to use your o-scope's Channel 1 volts/division setting to adjust the voltage to 20 mV.

(continues)

LAB ACTIVITY 5-2

(continued)

9   Set the function generator at a frequency of 5 kHz.

10   Have the instructor check your circuit.

11   Calculate and record the DC voltage drops and *static* voltage gain of the CE amplifier.

$$V_{R1} = \frac{(R_1)(V_{CC})}{R_1 + R_2} =$$

$$V_{R2} = \frac{(R_2)(V_{CC})}{R_1 + R_2} =$$

Check: $V_{R1} + V_{R2} = V_{CC} =$

$$V_E = V_{R2} - V_{BE} =$$

$$I_E = \frac{V_E}{R_4} =$$

$$V_{R3} = (I_C)(R_3) =$$

$$V_{CE} = V_{CC} - V_{R3} - V_E =$$

$$r'e = \frac{25mV}{I_E} =$$

$$A_V = \frac{R_3}{r'e} =$$

12   Let's measure and record the voltage dropped across the transistor, or V_{CE}. Set your DMM to DC volts. Connect the positive lead of your DMM to the collector of your transistor and the negative lead of your DMM to the emitter of the transistor. Power on *only* the 12 V DC voltage source and record the V_{CE}.

V_{CE} _____

How does this measured V_{CE} compare to the calculated V_{CE} in Step 11? If the V_{CE} is about one-half of the V_{CC} value, then the transistor is midpoint biased. We'll discuss this concept in detail in Chapter 6. Power off the DC voltage source and remove your DMM leads.

13   Power on the 12 V DC voltage source, the function generator, and the o-scope. Use your o-scope to ensure the AC input signal on Channel 1 is 20 mV peak-to-peak. You may have to adjust the Channel 1 volts/division setting and/or other o-scope settings. Record the value below.

V_{IN} _____

LAB ACTIVITY 5-2

14 Measure and record the AC output voltage that appears on the o-scope's Channel 2. You may have to adjust the Channel 2 volts/division setting and/or other o-scope settings. Record the value below.

V_{OUT} ————

15 Use the equation below to calculate and record the dynamic voltage gain of the CE amplifier.

$$A_V = \frac{V_{OUT}}{V_{IN}} =$$

A_V ————

How does this measured A_V compare to the calculated static A_V of Step 11? Explain any differences.

16 Set your o-scope's vertical mode switch to the dual setting. Adjust your o-scope settings so that the 20 mV input signal is on top and the V_{OUT} signal is directly below it. Are the waveforms mirror images; that is, does one peak positive at the same time the other peaks negative?

17 Power off the 12 V DC source, the function generator, and the o-scope. Remove capacitor C_B from the circuit. Power on the 12 V DC source, the function generator, and the o-scope. Use the equation below to recalculate and record the dynamic voltage gain of the CE amplifier.

$$A_V = \frac{V_{OUT}}{V_{IN}} =$$

A_V ————

How does this measured A_V compare to the measured A_V of Step 15? Explain any differences.

18 Have the instructor verify your results.

19 Build the circuit on Multisim®, Electronics Workbench®, or SPICE and then compare the results.

LAB ACTIVITY 5 - 3

Using Solid State Challenges™ to Troubleshoot Common-Emitter Amplifiers

Using the PC at your workstation, open the Solid State Challenges™ to the lesson "Troubleshooting Common-Emitter Amplifiers." Read the introductory screen "Troubleshooting Voltage-Divider Biased Amplifiers" and then click on the Begin tab. Read the directions in the red box, and then perform circuit checks to determine faults in common-emitter amplifiers.

5-4 COMMON-COLLECTOR AMPLIFIER

The common-collector (CC) amplifier is the second most widely used BJT amplifier configuration. Figure 5-7 shows a common-collector amplifier with a silicon transistor.

For a common-collector amplifier, the input signal is applied to the base and the output signal appears at the emitter. The common-collector is also called an **emitter follower**, or voltage follower, because the output voltage tracks or follows the amplitude and phase of the input voltage; that is, the output signal is reaching its positive peak when the input signal is reaching its positive peak. The common-collector has a low output impedance that makes it an excellent choice as the output amplifier stage for a radio receiver. The low output impedance provides a good match for speakers that typically have an input impedance of 4 Ω to 16 Ω. Remember, maximum transfer of power occurs between devices with the same impedances.

Another feature of the common-collector amplifier is its relatively low voltage gain and medium current gain. Because a common-collector has A_V of less than or equal to 1, it isn't used as a voltage amplifier. However, the common-collector has a midrange current gain, so it's typically used as a current amplifier.

Notice in Figure 5-7 that there's no collector resistor. The bypass capacitor C_B that appeared across the emitter resistor in the common-emitter

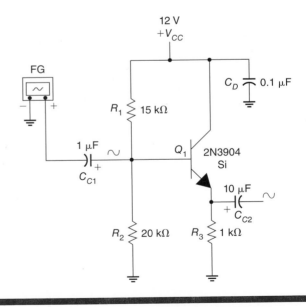

FIGURE 5-7 Common-collector amplifier with Si transistor (© Cengage Learning 2012)

amplifier has also been removed. For the common-collector amplifier, the capacitor C_D has been added in series with the DC power supply voltage V_{CC}. This capacitor is called a **decoupling capacitor**, and it acts like a bypass capacitor for the collector terminal. For the common-collector amplifier, all of the AC signal should appear at the emitter. If any of the AC signal leaks through the collector terminal, the job of capacitor C_D is to divert the AC away from V_{CC} and dump this AC to ground, where it will enter the base terminal through resistor R_2.

Let's go straight to Lab Activity 5-4 to see if a common-collector amplifier has a voltage gain of ≤ 1.

LAB ACTIVITY 5-4

Common-Collector Amplifier

Materials, Equipment, and Parts:

- *NTE* catalog or Internet access, www.nteinc.com (*Look it up to hook it up.*)

- PC w/Multisim®, Electronics Workbench®, or SPICE.

- DMM with test leads.

- Dual trace oscilloscope w/BNC-to-alligator leads.

- 12 V DC voltage source.

- Function generator, 1 Hz to 10 kHz.

- Breadboard and connecting wires.

- 2N3904 or equivalent NPN transistor.

- .1 µF capacitor, 1 µF electrolytic capacitor, and 10 µF electrolytic capacitor.

- 1 kΩ fixed resistor, 15 kΩ fixed resistor, and 20 kΩ fixed resistor.

Discussion Summary:

A transistor can be used as a common-collector amplifier. The CC has an AC input signal applied to the base terminal, and the AC output signal appears at the emitter terminal. One feature of the CC is its low output impedance. A more noticeable feature of the CC is that its voltage gain, or A_V, is ≤ 1, so the CC can't be used as a voltage amplifier. In this lab activity, we'll build a common-collector amplifier and prove that its A_V is @ 1.

Procedure:

SAFETY FIRST. Eye protection should always be worn when working with live voltages. Before powering on a live circuit, always check with your instructor.

1 Write the part number of the transistor at your workstation. Then use the *NTE* catalog or visit the website www.nteinc.com to find and record the NTE replacement number and the diagram number.

Part number _____

NTE number _____

Diagram # _____

2 Draw the transistor outline and label the emitter, base, and collector.

(continues)

LAB ACTIVITY 5-4

(continued)

3 Build the circuit of Figure 5-8 on your breadboard.

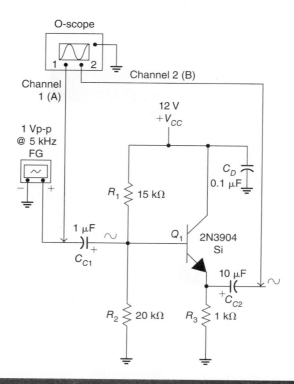

FIGURE 5-8 Common-collector amplifier with Si transistor (© Cengage Learning 2012)

4 Connect the positive lead of the 12 V DC voltage source to the top end of resistor R_1, the collector of Q_1, and capacitor C_D.

5 Connect the positive lead of the function generator to the negative side of capacitor C_{C1}.

6 Connect the positive lead of the oscilloscope's Channel 1 to the negative side of capacitor C_{C1}. Connect the positive lead of the o-scope's Channel 2 to the negative side of capacitor C_{C2}.

7 Connect the black lead of the function generator, the black lead of the 12 V DC voltage source, the black leads of the o-scope, a wire from the bottom of resistor R_2, a wire from the bottom of the emitter resistor R_3, and a wire from the bottom of capacitor C_D to one point (hole) on the breadboard, which is ground.

8 Set the function generator to the sine wave setting and to an amplitude of 1 V peak-to-peak. You may have to use your o-scope's Channel 1 volts/division setting to adjust the voltage to 1 V.

LAB ACTIVITY 5-4

9 Set the function generator at a frequency of 5 kHz.

10 Have the instructor check your circuit.

11 Power on your 12 V DC voltage source, the function generator, and the o-scope.

12 Use your o-scope to ensure the AC input signal on Channel 1 is 1 V peak-to-peak. You may have to adjust the Channel 1 volts/division setting and/or other o-scope settings. Record the value below.

V_{IN} _____

13 Measure and record the AC output voltage that appears on the o-scope's Channel 2. You may have to adjust the Channel 2 volts/division setting and/or other o-scope settings. Record the value below.

V_{OUT} _____

14 Use the equation below to calculate and record the dynamic voltage gain of the CC amplifier.

$$A_V = \frac{V_{OUT}}{V_{IN}} =$$

A_V _____

How does this measured A_V compare to the theoretical A_V of a CC? Explain any differences.

15 Have the instructor verify your results.

16 Build the circuit on Multisim®, Electronics Workbench®, or SPICE and then compare the results.

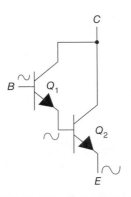

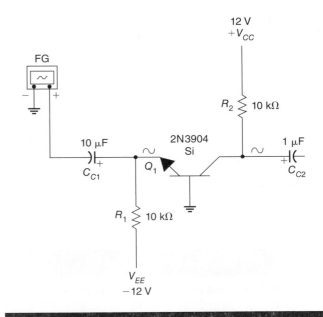

FIGURE 5-9 Darlington pair transistor (© Cengage Learning 2012)

Lab Activity 5-4 proved that a common-collector amplifier has a voltage gain of ≤ 1, which means that it can't be used to amplify a voltage signal. However, the common-collector has a midrange current gain, so it can be used as a current amplifier.

Sidney Darlington, an engineer at Bell Labs, increased the current amplification capability of the common-collector in 1953 by putting two common-collector transistors into one housing. Figure 5-9 shows the schematic symbol for the device known as the **Darlington amplifier or Darlington pair**.

The Darlington pair is really a two-stage amplifier. In Figure 5-9, you can see that the collectors of both NPN transistors are connected to each other and that the emitter of transistor Q_1 is connected to the base of Q_2. The overall current gain of the Darlington in Figure 5-9 is equal to the current gain of transistor Q_1 multiplied by the current gain of transistor Q_2. For example, if Q_1 has a current gain of 50 and Q_2 has a current gain of 40, then the overall current gain would be 50 times 40, or 2000.

Like most transistors, the Darlington amplifier is manufactured as an IC or as a discrete (individual) device with three leads. In fact, the only way to tell that a transistor is a Darlington pair is *Look it up to hook it up*.

5-5 COMMON-BASE AMPLIFIER

The common-base (CB) amplifier is the least common BJT amplifier configuration. Figure 5-10 shows a common-base amplifier with a silicon transistor.

For a common-base amplifier, the input signal is applied to the emitter terminal and the output signal

FIGURE 5-10 Common-base amplifier with Si transistor (© Cengage Learning 2012)

appears at the collector. The common-base has a low input impedance, which makes it an excellent choice as the input amplifier stage for a radio receiver. The common-base has a midrange A_V, which makes it suitable as a voltage amplifier. However, its current gain is ≤ 1, so it can't be used as a current amplifier.

Visually, the common-base amplifier circuit looks drastically different from the common-emitter and common-collector circuits. First, the transistor looks like it's positioned sideways. Second, the common-base has an additional negative supply voltage on the emitter terminal, called V_{EE}. The circuit of Figure 5-10 is an example of another type of biasing scheme for transistor amplifiers, called **emitter bias**. Emitter bias is used less often than voltage-divider bias because it requires two power supplies, one positive and one negative, and thus you have the added expense in the overall cost of a system. However, the common-base has a good voltage gain and is used in radio frequency receivers because its low input impedance doesn't negatively affect the incoming radio signals.

5-6 DECIBEL (dB) GAIN FOR AMPLIFIERS

Voltage gain A_V and power gain A_P are the two most common gains calculated and measured by technicians in the field. However, sometimes gains can be extremely high or extremely low, so another unit was developed to represent these extremes. The

decibel (dB) is a unit used to express the ratio of one unit to another. The two decibel equations below use the logarithm function for calculating gain. The logarithm (often called log) of a number is an exponent. Simply put, taking the log of a number reduces its value to a more manageable amount.

This concept is useful in the manufacture of meters. For example, labeling an analog meter—one with a needle or gauge—would be a challenge if you want to measure power gains from 2 to 10,000. How would you mark the divisions? With decibels, you can represent the same range as 3 dB to 40 dB. Let's see how.

EQUATION 5-6

Amplifier power gain in decibels

$$A_P\,(dB) = 10 \log \frac{P_{OUT}}{P_{IN}}$$

EXAMPLE 5

Situation

An amplifier has an output power of 2 W and an input power of 200 µW. What is the power gain? What is the power gain in decibels?

Solution

First, we'll find the power gain using Equation 5-1.

$$A_P = \frac{P_{OUT}}{P_{IN}} = \frac{2\,W}{200\,\mu W}$$

$$= 10,000$$

Now, let's use the dB power gain equation. To use this equation, go from right to left. That is, calculate the power gain, then press log, then press ×10.

$$A_P\,(dB) = 10 \log \frac{P_{OUT}}{P_{IN}}$$

$$= 10 \log 10000 = 40\,dB$$

For Example 5-5, the output power is 10,000 times greater than the input power or a change of +40 dB. Do you see how dB makes the quantity smaller? Just like gain, dB is just a number. Decibel values are also useful for troubleshooting communication systems such as cable or satellite TV. Let's see another example.

EXAMPLE 6

Situation

An amplifier has an output power of 70 W and an input power of 160 W. What is the power gain? What is the power gain in decibels?

Solution

First, we'll find the power gain using Equation 5-1.

$$A_P = \frac{P_{OUT}}{P_{IN}} = \frac{70\,W}{160\,W} = 0.4375$$

Now, let's use the dB power gain equation. To use this equation, go from right to left. That is, calculate the power gain, then press log, then press ×10.

$$A_P\,(dB) = 10 \log \frac{P_{OUT}}{P_{IN}}$$

$$= 10 \log .4375 = -3.59\,dB$$

For Example 5-6, the output power is 0.4375 times greater than the input power or a change of −3.59 dB. This means it is a power loss since you have less power coming out than going in. *Negative dB is a power loss.* Let's look at the dB equation for voltage gain (or loss).

EQUATION 5-7

Amplifier voltage gain in decibels

$$A_V\,(dB) = 20 \log \frac{V_{OUT}}{V_{IN}}$$

Notice the equation for dB voltage gain uses the number 20.

EXAMPLE 7

Situation

An amplifier has an output voltage of 6.5 V and an input voltage of 15 mV. What is the voltage gain? What is the voltage gain in decibels?

Solution

First, we'll find the voltage gain using Equation 5-3.

$$A_V = \frac{V_{OUT}}{V_{IN}} = \frac{6.5\,V}{15\,mV}$$

$$= 433.33$$

Now, let's use the dB voltage gain equation. Again, to use this equation, go from right to left. That is, calculate the voltage gain, then press log, then press ×20.

$$A_V\,(dB) = 20\,\log \frac{V_{OUT}}{V_{IN}}$$

$$= 20\,\log 433.33$$

$$= 52.73\,dB$$

EXAMPLE 8

Situation

An amplifier has an output voltage of 9 V and an input voltage of 12 V. What is the voltage gain? What is the voltage gain in decibels?

Solution

First, we'll find the voltage gain using Equation 5-3.

$$A_V = \frac{V_{OUT}}{V_{IN}} = \frac{9\,V}{12\,V} = 0.75$$

Now, let's use the dB voltage gain equation. Again, to use this equation, go from right to left. That is, calculate the voltage gain, then press log, then press ×20.

$$A_V\,(dB) = 20\,\log \frac{V_{OUT}}{V_{IN}}$$

$$= 20\,\log 0.75$$

$$= -2.49\,dB$$

For Example 5-7, the output voltage is 433.33 times greater than the input voltage or a change of +52.73 dB. Again, see how dB makes the quantity smaller? Just like gain, dB is just a number. Let's do another.

For Example 5-8, the output voltage is 0.75 times greater than the input voltage or a change of −2.49, which means the amplifier has a voltage loss. Again, a negative dB represents a loss. The

calculation and understanding of dB can help a technician in troubleshooting amplifier systems.

In Section 5-3, we mentioned that many electronic systems, such as a radio receiver, have multiple amplifier stages. Figure 5-11 is a block diagram of a simple radio receiver.

The radio signal enters through the antenna on the far left and produces sound from the speaker on the far right. The CB amplifier is used at the

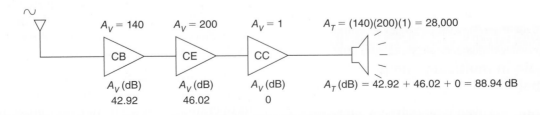

FIGURE 5-11 Block diagram of a simple radio receiver (© Cengage Learning 2012)

input stage because it offers low impedance, or opposition, to the incoming radio signal. The CE amplifier is in the center stage because it provides a midrange voltage gain and a high power gain. The CC amplifier is at the output stage because it has a low output impedance that will match the low input impedance of the speaker.

We know that the CB amplifier and the CE amplifier stages have a midrange voltage gain and that the CC amplifier has an A_V of ≤ 1. To figure the total gain for all stages, multiply the gain for each stage.

EQUATION 5-8

Total gain in multistage amplifier

$$A_T = (A_1)(A_2)(A_3)$$

EXAMPLE 9

Situation

A multistage amplifier has a CB amplifier with an A_V of 140, a CE amplifier with an A_V of 200, and a CC with an A_V of 1. What is the overall voltage gain of the three stages?

Solution

$$A_T = (A_1)(A_2)(A_3)$$
$$= (140)(200)(1) = 28,000$$

If you use dB values, the equation changes for calculating total gain in a multistage amplifier. In the decibel equation, you simply add the stage gains together.

EQUATION 5-9

Total gain in multistage amplifier in decibels

$$A_T(dB) = A_1(dB) + A_2(dB) + A_3(dB)$$

EXAMPLE 10

Situation

A multistage amplifier has a CB amplifier with an A_V of 42.92 dB, a CE amplifier with an A_V of 46.02 dB, and a CC with an A_V of 0 dB. What is the overall voltage gain of the three stages in decibels?

Solution

$$A_T (dB) = A_1 (dB) + A_2 (dB) + A_3 (dB)$$
$$= 42.92\,dB + 46.02\,dB + 0\,dB$$
$$= 88.94\,dB$$

If you do the math, you'll see that we just used the voltage gains of the stages from Example 5-9.

Thus, it's easier to determine the total gain for multistage amplifiers when using decibels because you just add up the gain of all the stages. The same equations can be applied to power gains and current gains. This is another example where decibels values come in handy.

CHAPTER SUMMARY

The transistor is a three-terminal semiconductor device that can be used as a high-speed switch or as an amplifier. When used as an amplifier, the transistor takes a small signal (usually AC) and makes it bigger. The generic schematic symbol for an amplifier is a triangle with one of its points facing to the right. The three most important characteristics of an amplifier are input impedance (Z_{IN}), output impedance (Z_{OUT}), and gain.

The gain of an amplifier is simply the output signal divided by the input signal; that is, how many times bigger the output signal is than the input signal. Amplifiers can have three types of gain: power gain (A_P), current gain (A_I), and voltage gain (A_V).

A transistor amplifier usually amplifies an AC signal, but first it needs a DC voltage (bias) to supply its operating power. The most common method used to bias transistor amplifiers is called voltage-divider bias, where the main DC power supply voltage is divided among various resistors to turn on a transistor so that it's ready for amplification.

The three transistor amplifier configurations are the common-emitter, the common-collector, and the common-base. A transistor amplifier configuration is determined by the relationship between the AC input signal and the AC output signal. The common-emitter has the input signal applied to the base terminal and has the output signal appear at the collector terminal. If the input signal is applied to the base and the output signal appears at the emitter, then the configuration is called a common-collector. If the input signal is applied to the emitter and the output signal appears at the collector, then the configuration is called a common-base. Each configuration has particular values of Z_{IN}, Z_{OUT}, and gain.

The DC voltage drops and A_V for amplifiers can be calculated and measured. The static gain is determined with no AC signal applied to the amplifier, while the dynamic gain is determined with an AC signal applied.

The decibel power and voltage gains of amplifiers can also be calculated. Decibels reduce a gain to a smaller value, and they are useful for troubleshooting communication systems such as cable or satellite TV.

Many electronic systems such as a radio receiver have multiple amplifier stages. To calculate the overall gain, the individual stages gains are multiplied together. If the gains are in dB values, you simply add the individual stage gains together.

CHAPTER EQUATIONS

Amplifier power gain

(Equation 5-1)

$$A_P = \frac{P_{OUT}}{P_{IN}}$$

Amplifier current gain

(Equation 5-2)

$$A_I = \frac{I_{OUT}}{I_{IN}}$$

Amplifier voltage gain

(Equation 5-3)

$$A_V = \frac{V_{OUT}}{V_{IN}}$$

Amplifier AC emitter resistance

(Equation 5-4)

$$r'e = \frac{25\,mV}{I_E}$$

Amplifier static voltage gain

(Equation 5-5)

$$A_V = \frac{R_3}{r'e}$$

Amplifier power gain in decibels

(Equation 5-6)

$$A_P\,(dB) = 10\log\frac{P_{OUT}}{P_{IN}}$$

Amplifier voltage gain in decibels

(Equation 5-7)

$$A_V\,(dB) = 20\log\frac{V_{OUT}}{V_{IN}}$$

Total gain in multistage amplifier

(Equation 5-8)

$$A_T = (A_1)(A_2)(A_3)$$

Total gain in multistage amplifier in decibels

(Equation 5-9)

$$A_T(dB) = A_1(dB) + A_2(dB) + A_3(dB)$$

CHAPTER REVIEW QUESTIONS

Chapter 5-1

1. What is an amplifier?

2. Draw and label the generic schematic symbol for an amplifier.

3. Why must the impedances of external devices connected to an amplifier match the amplifier's Z_{IN} and Z_{OUT}?

4. An amplifier has an output power of 17 W and an input power of 90 mW. What is the power gain?

5. An amplifier has an output power of 30 W and an input power of 200 mW. What is the power gain?

6. An amplifier has an output current 80 mA and an input current of 400 µA. What is the current gain?

7. An amplifier has an output current 4.6 A and an input current of 33 mA. What is the current gain?

8. An amplifier has an output voltage of 5 V and an input voltage of 42 mV. What is the voltage gain?

9. An amplifier has an output voltage of 7.8 mV and an input voltage of 90 µV. What is the voltage gain?

10. What is the other name for bias?

Chapter 5-2

11. True or false. A transistor's operating power is DC, but a transistor amplifies an AC signal.

12. Using a DMM, how would you measure the V_{CE} of a transistor amplifier?

Chapter 5-3

13. How does one identify a transistor amplifier configuration?

14. Which amplifier configuration has a low output impedance?

15. Which amplifier configuration has a 180° phase shift between the input and output signal?

16. Which amplifier configuration has an A_I of ≤1?

17. What is the approximate voltage gain of the amplifier configuration of Figure 5-12?

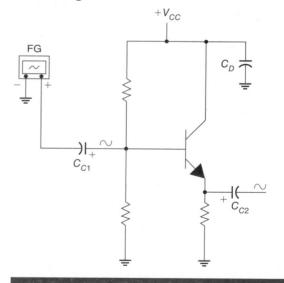

FIGURE 5-12　Amplifier configuration (© Cengage Learning 2012)

18. Which amplifier configuration is also known as an emitter follower?

19. **Which amplifier configuration is most commonly used?**

20. **How does an AC signal going into a CE amplifier acquire gain?**

21. **What function do coupling capacitors serve in a transistor amplifier?**

22. **What is the purpose of a bypass capacitor in a CE amplifier?**

23. **Determine the following values for the circuit of Figure 5-13.**

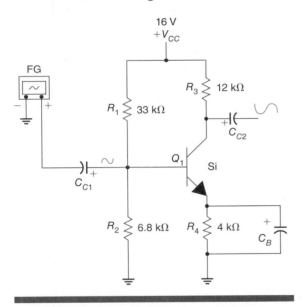

FIGURE 5-13 Common-emitter amplifier with Si transistor (© Cengage Learning 2012)

$$V_{R1} =$$

$$V_{R2} =$$

Check: $V_{R1} + V_{R2} = V_{CC} =$

$$V_E =$$

$$I_E =$$

$$V_{R3} =$$

$$V_{CE} =$$

$$r'e =$$

$$A_V =$$

24. **What is the first step in troubleshooting a transistor amplifier circuit?**

Chapter 5-4

25. **Name one application for a common-collector circuit.**

26. **What is the purpose of a decoupling capacitor in a CC amplifier?**

27. **Draw the schematic symbol for a Darlington amplifier.**

Chapter 5-5

28. **Common-base amplifiers often use emitter bias. Why is voltage-divider bias preferred to emitter bias for powering a transistor amplifier?**

29. **What is the current gain of a common-base amplifier?**

Chapter 5-6

30. **An amplifier has an output power of 4 W and an input power of 1 mW. What is the power gain? What is the power gain in decibels?**

31. **An amplifier has an output power of 90.5 W and an input power of 131 W. What is the power gain? What is the power gain in decibels?**

32. **An amplifier has an output voltage of 16 V and an input voltage of 45 mV.**

What is the voltage gain? What is the voltage gain in decibels?

33. An amplifier has an output voltage of 11 V and an input voltage of 22 V. What is the voltage gain? What is the voltage gain in decibels?

34. A multistage amplifier has a CB amplifier with an A_V of 180, a CE amplifier with an A_V of 250, and a CC with an A_V of 1. What is the overall voltage gain of the three stages?

35. A multistage amplifier has a CB amplifier with an A_V of 42.92 dB, a CE amplifier with an A_V of 46.02 dB, and a CC with an A_V of −1.25 dB. What is the overall voltage gain of the three stages in decibels?

Amplifier Classes

OBJECTIVES   *Upon completion of this chapter, you should be able to:*

- Define amplifier fidelity, distortion, and efficiency.
- Calculate an amplifier's efficiency.
- Distinguish among transistor cutoff, midpoint bias, and saturation.
- Identify and distinguish the classes of amplifiers.
- Build a class B amplifier and identify crossover distortion.

- Build a class AB amplifier to eliminate crossover distortion.
- Calculate the resonant frequency of the tank section of a class C amplifier.
- Build a class C amplifier to calculate and measure its resonant frequency and dynamic gain.

MATERIALS, EQUIPMENT, AND PARTS

Materials, equipment, and parts needed for the lab experiments in this chapter are listed below:

- *NTE* catalog or Internet access, www.nteinc.com
- PC w/Multisim®, Electronics Workbench®, or SPICE.
- DMM with test leads.
- Dual trace oscilloscope w/BNC-to-alligator leads.
- 12 V DC voltage source.
- Function generator, 1 Hz to 30 kHz.

- Breadboard and connecting wires.
- 2N3904 or equivalent NPN transistor.
- 2N3906 or equivalent PNP transistor.
- 1N4148 diode (2).
- 680 pF capacitor, 1 µF electrolytic capacitor (2), 10 µF

electrolytic capacitor (2), and 220 µF electrolytic capacitor.
- 47 mH inductor coil.
- 33 Ω fixed resistor, 180 Ω fixed resistor, 1 kΩ fixed resistor, 1.8 kΩ fixed resistor (2), and 100 kΩ fixed resistor.

GLOSSARY OF TERMS

Amplifier class The category of amplifier that is determined by the way a transistor is biased; that is, the various DC voltages that affect the transistor's operation

Fidelity The ability of an amplifier to "faithfully," or accurately, reproduce an AC input signal

Distortion The unwanted change in the shape of an AC signal

Efficiency The percentage of DC power that actually makes it to the amplifier's load, such as another amplifier stage or a speaker; result of AC output power divided by the DC input power

Class A amplifier An amplifier that is biased at midpoint and has very high fidelity and low efficiency

Cutoff The operating condition of a bipolar junction transistor in which the transistor has no I_B, no I_C, and is totally off

Saturation The operating condition of a bipolar junction transistor where the transistor is totally on, the I_C is at maximum, and increasing the I_B won't increase the I_C anymore

Midpoint operation The most desirable operating condition for a transistor amplifier, occurring

when the I_C is about one-half the current value at saturation and V_{CE} is about one-half the value of V_{CC}

Small signal amplifier A transistor circuit that amplifies voltage signals in the µV to mV range and up to about 0.5 W of power; these small voltages are found at the input of radio receivers, cassette tape heads, vinyl record players, and the preamplifier stage for guitar amplifiers

Class B amplifier An amplifier that is biased at cutoff and has high fidelity and high efficiency; it usually contains an NPN transistor that reproduces the positive alternation of the AC input signal, and a PNP transistor that reproduces the negative alternation of the AC input signal

Push-pull An amplifier circuit that uses two transistors working together, one conducting only during the positive alternation of an AC input signal, and the other conducting only during the negative alternation of an AC input signal

Crossover distortion The undesirable changing of an amplified signal in a class B amplifier that occurs as the task

of reproducing the AC input signal is passed on, or "crosses over," from one transistor to the next; it is shown visually by a small horizontal line, or "glitch," between the positive and negative alternations of the AC output signal

Compensating diode(s) Diodes used in a class AB amplifier that each provide a voltage drop across the base-to-emitter junction of a transistor to keep it barely on and thus eliminate crossover distortion

Large signal amplifier A transistor circuit that provides a power greater than 0.5 W; used in class B and AB amplifiers as the intermediate stage or output stage for audio applications

Class C amplifier An amplifier that is biased way below cutoff and has low fidelity and excellent efficiency; usually contains a tank section connected to the transistor's collector that creates an alternating waveform, restoring most of the distorted AC waveform

Resonant frequency (f_r) The frequency of a tuned circuit in an amplifier that is used at the input of radio receivers or at the output of radio transmitters

6-1 AMPLIFIER FIDELITY VERSUS EFFICIENCY

In Chapter 5, we learned that a transistor amplifier configuration—CE, CC, or CB—is determined by the relationship between the AC input signal and the AC output signal. We also learned that each configuration has unique values of Z_{IN}, Z_{OUT}, and gain.

In addition to having a configuration, every transistor amplifier belongs to a particular class. An **amplifier class** is determined by the way an amplifier is biased; that is, the various DC voltages that affect the transistor's operation. This chapter focuses on the four main amplifier classes—A, AB, B, and C—and their unique values of fidelity and efficiency.

INTERNET ALERT

To learn about additional amplifier classes not covered in this chapter, check the website http://en.wikipedia.org/wiki/Electronic_amplifier#Power_amplifier_classes

The **fidelity** of an amplifier is its ability to "faithfully," or accurately, reproduce an AC input signal. For example, if you play a music CD on a high-fidelity stereo system, the quality of sound will make you feel like you're in the recording studio with the artist. On the other hand, a stereo system with poor fidelity will result in music that misses the high and low frequencies; that is, music that sounds unclear or distorted. **Distortion** is an unwanted change in the shape of an AC signal. Thus, high fidelity means little or no distortion.

Remember from Chapter 5 that an amplifier doesn't make an AC signal larger by magic. The AC signal steals power from the V_{CC} (DC supply voltage) to get larger. The **efficiency** of an amplifier is the percentage of DC power that actually makes it to the amplifier's load, such as another amplifier stage or a speaker. For example, say an

amplifier is attached to a speaker. If the amplifier has an efficiency of 70%, then 70% of the V_{CC} power will go to the speaker; the other 30% is involved in making the AC signal larger.

EQUATION 6-1

Amplifier efficiency

$$\eta = \frac{P_{AC}}{P_{DC}}(100)$$

η = % efficiency
P_{AC} = AC output signal power
P_{DC} = DC input power (V_{CC})

EXAMPLE 1

Situation

An amplifier has an AC output power of 300 mW and a DC input power of 1.25 W. What is the efficiency of the amplifier?

Solution

$$\eta = \frac{P_{AC}}{P_{DC}}(100) = \frac{300\,mW}{1.25\,W}(100)$$

$$= 0.24(100)$$

$$= 24\%$$

An efficiency of 24% is considered poor because only 24% of the power provided by the V_{CC} actually makes it to whatever load the amplifier is driving, such as another amplifier stage or a speaker. The other 76% is being "wasted" by the amplifier. In theory, the maximum efficiency of an amplifier is 100%. However, the transistor itself consumes power in the form of heat, so an amplifier couldn't possibly give all its DC power to its load. On the low end, an efficiency of 0% means the transistor is using all the available DC power for itself and is not delivering any power to the load.

EXAMPLE 2

Situation

An amplifier has a DC input power of 105 W and an AC output power of 72 W. What is the efficiency of the amplifier?

Solution

$$\eta = \frac{P_{AC}}{P_{DC}}(100) = \frac{72\,W}{105\,W}(100)$$

$$= 0.6857\,(100)$$

$$= 68.57\%$$

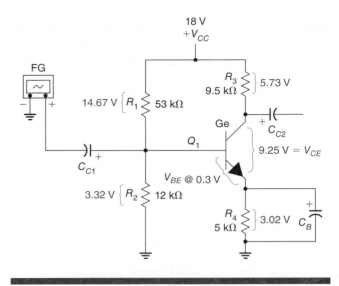

FIGURE 6-1   Class A amplifier with midpoint biasing (© Cengage Learning 2012)

An efficiency of 68.57% is better than average. The amplifier uses only 31.43% of the power from the V_{CC}; thus, it delivers 68.57% to its load, which might be a speaker or other output device.

The different amplifier classes have varying values of fidelity and efficiency. Table 6-1 shows the different classes of amplifiers.

Notice in Table 6-1 that there's usually a trade-off: an amplifier can't have both excellent fidelity *and* excellent efficiency. In the upcoming sections, we'll look at each amplifier class and investigate its strengths and weaknesses.

6-2 CLASS A AMPLIFIERS

The **class A amplifier** is an amplifier that is biased at midpoint and has very high fidelity and low efficiency. According to Table 6-1, the class A amplifier has very high fidelity; that is, the AC output signal of the class A amplifier will be a larger but exact duplicate of the AC input signal. The class A amplifier also has an efficiency of only @ 25%, which is considered poor. Both fidelity and efficiency are determined by the biasing of the transistor: the way the main DC power supply voltage (V_{CC}) is divided among various resistors to turn on a transistor so that it's ready for amplification. Figure 6-1 shows a class A

TABLE 6-1   Amplifier Classes (way it is biased) (© Cengage Learning 2012)

CLASS	BIASED	CONDUCTION ANGLE	FIDELITY	EFFICIENCY	SPECIAL FEATURES
A	Midpoint	360°	Very high	Up to 25%	Small signal amplifiers (especially audio)
B	Cutoff	180° (per transistor)	High (subject to crossover distortion)	Up to 78.5%	RF and audio amplifiers; push-pull
C	Below Cutoff	<180°	Low	Up to 99%	RF power amplifiers; contains tank circuit

amplifier that uses voltage-divider bias to provide midpoint biasing.

The transistor in Figure 6-1 is midpoint biased. What exactly does that mean? To understand the midpoint of something, we have to look at the lower and upper limits. Remember, biases are DC voltages dropped across the various terminals of a transistor when it is static—not yet amplifying an AC signal. We calculated the static voltages and currents of common-emitter amplifiers in Chapter 5. If a transistor has a DC supply voltage, or V_{CC}, applied to it, then the transistor is consuming power in the form of voltage drops and currents.

A transistor has three basic possible operating conditions: cutoff, saturation, and midpoint operation. At **cutoff**, the transistor has no I_B and thus no I_C; it is acting like an open switch, so the voltage dropped across the transistor (V_{CE}) is maximum or equal to the V_{CC}. The transistor is totally off and at its lower limit of operation.

At **saturation**, the transistor is totally on. Increasing the I_B won't increase the I_C, and that's why it's called saturation. I_C is maximum at saturation, and the V_{CE} is at its minimum or 0 V because the transistor is acting like a closed switch. A closed switch has nearly 0 Ω resistance, and using Ohm's Law (V = IR), we see that V = 0 V. Saturation is the upper limit of operation for a transistor. When a transistor exceeds saturation, the distortion is high, and the transistor can overheat and burn up. Driving a transistor beyond saturation can also cut off or clip the positive and negative alternations of the AC input waveform; that is, distort the waveform.

Between these two limits is **midpoint operation**. When a transistor is midpoint biased, the I_C is about one-half the current value at saturation, and V_{CE} is about one-half the value of V_{CC}. Midpoint operation is the most desirable operating condition for a transistor amplifier. At midpoint operation, the transistor will amplify an incoming AC signal without clipping any portion of the signal, and thus distortion will be at a minimum. The three transistor operating conditions are summarized below:

Saturation: totally on, I_C = maximum, V_{CE} = 0 V.

Midpoint: $I_C \approx$ ½ maximum, $V_{CE} \approx$ ½ maximum (V_{CC}).

Cutoff: I_C = 0 A, V_{CE} = maximum (V_{CC}).

Looking at Figure 6-1, you can see that the V_{CE} is equal to 9.25 V, which is about one-half of the V_{CC}'s value of 18 V. Thus, the transistor in

Figure 6-1 is midpoint biased, and that's why it's a class A amplifier.

You can't just look at an amplifier and tell that it's a class A amplifier. Two methods are used to determine if an amplifier is class A: static and dynamic. Using the first method, you have to calculate the static voltage drops and currents and determine the voltage dropped across the collector-to-emitter, the V_{CE}. If the V_{CE} is about one-half of the V_{CC}, then the amplifier is a class A, and it has the features shown in the top row of Table 6-1.

The second method is performed by measuring V_{CE} with power applied. Set your DMM to the DC volts setting. Connect the positive lead of your DMM to the collector terminal and the negative lead of your DMM to the emitter terminal. If the measured DC voltage is about equal to one-half of the V_{CC}, then the amplifier is class A, and it has the features shown in the top row of Table 6-1.

Since the transistor of Figure 6-1 is midpoint biased, its current with no AC signal applied is about one-half of the maximum current. The transistor is wasting power because it's on all the time. This power being used by the transistor is not available to amplify an incoming AC signal. If we look again at Equation 6-1, $\eta = \dfrac{P_{AC}}{P_{DC}} (100)$, we see that the higher the value of DC power consumed, the lower the efficiency. This is why a class A amplifier has poor efficiency: it's on all the time, even when an AC signal is not present at its input.

However, the fact that the class A amplifier is on all the time does give it an advantage over other amplifiers. When the AC input signal arrives at the input of the amplifier, the transistor is ready for it, and the entire AC signal (all 360° of it) is reproduced. This is why a class A amplifier has very high fidelity: it's always ready to reproduce an incoming AC signal.

Midpoint biasing is the best choice for amplifying AC signals because of the very high fidelity. Class A amplifiers are used extensively in audio applications (20 Hz to 20,000 Hz) for amplifying sound with quality reproduction. However, since the class A amplifier has low efficiency, its use is usually limited to small signal amplifiers. **Small signal amplifiers** typically involve voltage signals in the μV to mV range and up to about 0.5 W of power. These small voltages are found at the input of radio receivers, cassette tape heads, vinyl record players, and the preamplifier stage for guitar amplifiers. As a result, class A amplifiers are used as the input stage in a multistage amplifier system. Since the common-emitter amplifier we built in Lab Activity 5-2 was also a class A amplifier (for Step 12, the measured V_{CE} was about one-half of the V_{CC}), there's no need to duplicate the lab here.

Here we have an important point: an amplifier is a certain configuration *and* a certain class at the same time. Remember, the configuration of an amplifier is determined by the relationship between the input signal and the output signal. The class of an amplifier is determined by the way it is biased. Thus, one transistor can be both a CE amplifier and a class A amplifier at the same time.

6-3 CLASS B AMPLIFIERS

The **class B amplifier** is an amplifier that is biased at cutoff and has high fidelity and high efficiency. According to Table 6-1, the class B amplifier is biased at cutoff, which means that it's off without an AC signal applied to its input ($I_C = 0$, $V_{CE} = V_{CC}$). Thus, the class B amplifier draws little current from the V_{CC}; the result is a high efficiency of nearly 80%. If the class B amplifier had just one transistor, by the time it turned on it would miss a good part of the AC signal. Thus, the output signal wouldn't look at all like the input signal: the signal would be distorted.

In practice, however, the class B has two transistors—one NPN and one PNP—biased at cutoff. The NPN transistor reproduces the positive alternation of the AC input signal, and the PNP reproduces the negative alternation of the AC input signal. Figure 6-2 shows a class B amplifier.

Notice that the AC input signal enters the base of both transistors and exits at their emitters; thus, both *configurations* are common-collector. Remember, an amplifier is both a class and a configuration. Because the class B amplifier uses common-collector configurations, it will have no voltage gain, but it will have midrange current and power gains.

The two transistors in Figure 6-2 look like they're stacked together, Q_1 on the top and Q_2 on the bottom: this is the key visual feature of the class B amplifier. Both of these transistors are made of silicon, so neither one will turn on until @ 0.6 V to 0.7 V is across its base-to-emitter junction (V_{BE}).

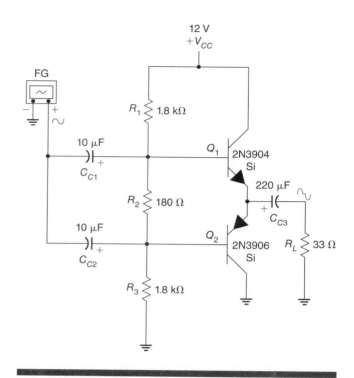

FIGURE 6-2   Class B amplifier with two transistors (© Cengage Learning 2012)

Likewise, they won't turn off until their V_{BE} drops below 0.6 V. During the positive alternation of the AC input signal, the NPN transistor Q_1 is on, and the PNP transistor Q_2 is off. During the negative alternation of the AC input signal, Q_2 is on, and Q_1 is off. Thus, the transistors aren't on at the same time. This arrangement is called **push-pull** because they take turns conducting.

A delay exists between the time it takes for Q_1 to be totally off and the time it takes for Q_2 to be totally on and vice versa. This delay affects the shape of the AC output signal, and it's called **crossover distortion**. Figure 6-3 shows an example of crossover distortion.

Crossover distortion occurs as the job of reproducing the AC input signal is passed on or crosses over from one transistor to the next. The crossover distortion is shown by a small horizontal line, or "glitch," between the positive and negative alternations of the AC output signal. Crossover distortion is undesirable because it changes the shape of the waveform. This is a trade-off for the class B amplifier; it can provide a midrange power gain, but its fidelity is lower than a class A amplifier because of the crossover distortion. The next section shows how to correct crossover distortion.

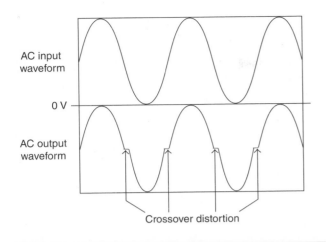

FIGURE 6-3 Crossover distortion in a class B amplifier (© Cengage Learning 2012)

6-4 CLASS AB
AMPLIFIERS

The class AB amplifier is really just a variation of the class B amplifier. Figure 6-4 shows a class AB amplifier and the method used to eliminate crossover distortion.

Notice in Figure 6-4 that diodes D_1 and D_2 have been inserted between the bases of Q_1 and Q_2. These silicon diodes are called **compensating diodes** because they are used to eliminate crossover distortion. D_1 provides a voltage drop of @ 0.6 V to 0.7 V across the base-to-emitter junction of Q_1 so that it is barely on. Likewise, D_2 provides a voltage drop of @ 0.6 V to 0.7 V across the base-to-emitter junction of Q_2 so that it is barely on. Thus, when the AC input signal changes from positive to negative, transistor Q_2 is already on and ready to reproduce the entire negative alternation. On the other hand, when the AC input signal goes from negative to positive, transistor Q_1 is already on and ready to reproduce the entire positive alternation.

The class AB amplifier has fidelity and efficiency values between those of a class A amplifier and those of a class B amplifier. Because class B

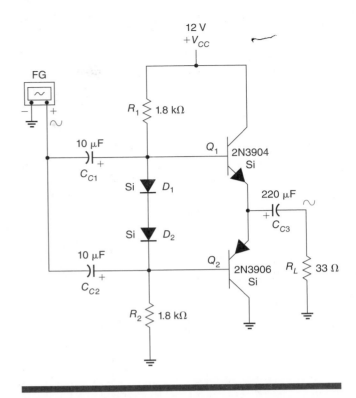

FIGURE 6-4 Eliminating crossover distortion in a class AB amplifier (© Cengage Learning 2012)

amplifiers and class AB amplifiers have high efficiency, they can be used as **large signal amplifiers**, which provide power greater than 0.5 W. Class B and AB amplifiers are usually used as the intermediate stage or output stage for audio applications. For example, class B and AB amplifiers can be used in radios for the audio output stage that is used to drive the speakers. They can also be used for the output stage in smaller battery-operated devices such as iPods or MP3 players.

The class B and AB amplifiers can be made of discrete devices like those shown in Figures 6-2 and 6-4. In many cases, however, audio amplifier integrated circuits are used. For example, the LM386 is an audio amplifier IC used in power applications up to 1 watt.

Let's build the class B and class AB amplifiers and test them live.

LAB ACTIVITY 6-1

Class B Amplifier

Materials, Equipment, and Parts:

- *NTE* catalog or Internet access, www.nteinc. com (*Look it up to hook it up.*)

- PC w/Multisim®, Electronics Workbench®, or SPICE.

- DMM with test leads.

- Dual trace oscilloscope w/BNC-to-alligator leads.

- 12 V DC voltage source.

- Function generator, 1 Hz to 10 kHz.

- Breadboard and connecting wires.

- 2N3904 or equivalent NPN transistor.

- 2N3906 or equivalent PNP transistor.

- 1N4148 diode (2).

- 10 µF electrolytic capacitor (2); 220 µF electrolytic capacitor.

- 33 Ω fixed resistor, 180 Ω fixed resistor, and 1.8 kΩ fixed resistor (2).

Discussion Summary:

The class B amplifier uses two transistors: one to amplify the positive alternation of the AC input signal and one to amplify the negative alternation. The class B amplifier configuration uses a common-collector configuration, so the output signal appears at the emitters of both transistors. The output signal should also have a glitch that shows the pause as one transistor stops conducting, and the other transistor starts conducting (crossover distortion). In this lab, we'll also modify the class B amplifier to eliminate the crossover; that is, make it a class AB amplifier.

Procedure:

SAFETY FIRST. Eye protection should always be worn when working with live voltages. Before powering on a live circuit, always check with your instructor.

1 Write the part number of the transistors at your workstation. Then use the *NTE* catalog or visit the website www.nteinc.com to find and record the NTE replacement numbers and the diagram numbers.

1	2
Part number _____	Part number _____
NTE number _____	NTE number _____
Diagram # _____	Diagram # _____

2 Draw the transistor outlines and label the emitter, base, and collector for each transistor.

1   2

(continues)

LAB ACTIVITY 6-1

(continued)

3   Build the circuit of Figure 6-5 on your breadboard.

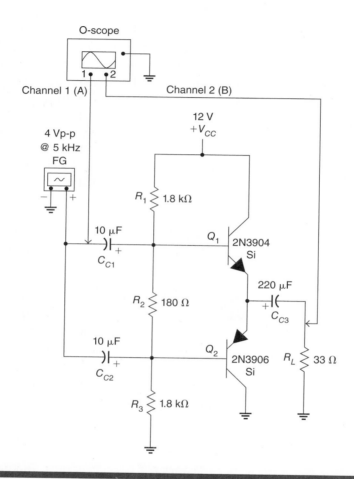

FIGURE 6-5   Class B amplifier (© Cengage Learning 2012)

4   Connect the positive lead of the 12 V DC voltage source to the top end of resistors R₁ and the collector of Q₁.

5   Connect the positive lead of the function generator to the negative side of capacitors C_C1 and C_C2.

LAB ACTIVITY 6-1

6 Connect the positive lead of the oscilloscope's Channel 1 to the negative side of capacitors C_{C1} and C_{C2}.

7 Connect the positive lead of the oscilloscope's Channel 2 to the positive or top end of the 33 Ω load resistor R_L.

8 Connect the black lead of the function generator, the black lead of the 12 V DC voltage source, the black leads of the o-scope, a wire from the bottom of resistor R_3, a wire from the collector of transistor Q_2, and a wire from the bottom of the 33 Ω load resistor R_L to one point (hole) on the breadboard, which is ground.

9 Set the function generator to the sine wave setting and to an amplitude of 4 V peak-to-peak. You may have to use your o-scope's Channel 1 volts/division setting to adjust the voltage to 4 V.

10 Set the function generator to a frequency of 5 kHz.

11 Have the instructor check your circuit.

12 Power on your 12 V DC voltage source, the function generator, and the o-scope.

13 Set your o-scope mode switch to dual. Adjust your oscilloscope volts/division settings and time/division settings to get the AC input waveform and AC output waveform to appear like those of Figure 6-3; that is, the AC input waveform should be on the top and the AC output waveform directly below it. The AC output waveform should have crossover distortion.

(continues)

LAB ACTIVITY 6-1

(continued)

14 Power off the 12 V DC source, the function generator, and the o-scope. Remove resistor R_2 (the 180 Ω resistor) from the circuit. Insert the 1N4148 diodes D_1 and D_2 in place of resistor R_2 as shown in Figure 6-6. Now, we've changed the class B amplifier to a class AB amplifier.

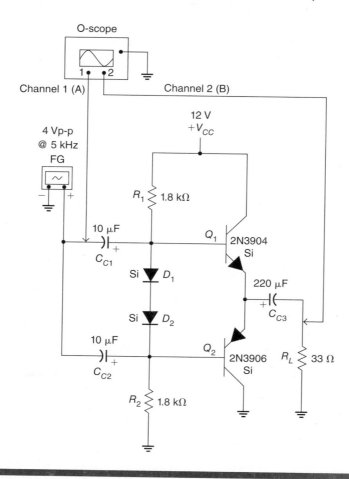

FIGURE 6-6   Class AB amplifier (© Cengage Learning 2012)

15 Power on the 12 V DC source, the function generator, and the o-scope. Has the AC output waveform changed? Did the two compensating diodes eliminate the crossover distortion?

16 Have the instructor verify your results.

17 Build the circuit on Multisim®, Electronics Workbench®, or SPICE and then compare the results.

6-5 CLASS C AMPLIFIERS

The **class C amplifier** is an amplifier that is biased way below cutoff and has low fidelity and excellent efficiency. Figure 6-7 shows a class C amplifier.

Notice that the AC input signal enters the base of transistor Q_1 and exits at the collector; thus, the *configuration* is a common-emitter. Remember, an amplifier is both a class and a configuration. Because the class C amplifier uses a common-emitter configuration, it will have midrange voltage and current gains and a high power gain.

The class C amplifier has poor fidelity because it's hardly on (<180° of the AC input signal). The AC input signal actually turns on the class C amplifier. By the time the amplifier starts conducting, most of the AC input signal has already passed through, so the amplifier only reproduces a small part of the AC input signal. Thus, the AC waveform is badly distorted: the output waveform looks very different from the input waveform. Because of its poor fidelity, the class C amplifier is rarely used in audio applications (frequencies below 20 kHz).

In Figure 6-7, notice the tank section connected to the transistor's collector. The tank section is the key visual feature of the class C amplifier. The capacitor in the tank circuit charges and discharges through the coil, creating an alternating waveform that restores most of the distorted AC waveform.

Link to Prior Learning

A tank section is composed of a coil and capacitor in parallel. The capacitor can be a fixed type or a variable type. If the capacitor is variable, it can be adjusted to tune the section to a desired frequency, called the **resonant frequency (f_r)**. A tank section with an adjustable capacitor is called a tuned circuit. Tuned circuits are used at the input of radio receivers or at the output of radio transmitters.

Since the tank section is critical to the operation of a class C amplifier, we'll review the equation used to calculate the resonant frequency of a tank section.

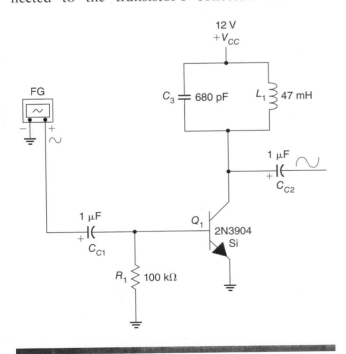

FIGURE 6-7   Class C amplifier (© Cengage Learning 2012)

EQUATION 6-2

Class C amplifier resonant frequency of tank section

$$f_r = \frac{1}{2\pi\sqrt{LC}}$$

f_r = resonant frequency in Hz

L = inductance of coil in tank

C = capacitance of capacitor in tank

When using the resonant frequency equation, work from right to left and bottom to top. For example, first multiply the coil and capacitor values, take the square root of this answer, multiply the new answer by 2π (6.28), and then use the invert function on your calculator.

EXAMPLE 3

Situation

What is the resonant frequency of a class C amplifier whose tank section has a coil measuring 47 mH and a capacitor measuring 220 pF?

Solution

$$f_r = \frac{1}{2\pi\sqrt{LC}}$$

$$= \frac{1}{2\pi\sqrt{(47\ mH)(220\ pF)}}$$

$$= 49{,}519.93\ Hz$$

EXAMPLE 4

Situation

What is the resonant frequency of a class C amplifier whose tank section has a coil measuring 100 mH and a capacitor measuring 470 pF?

Solution

$$f_r = \frac{1}{2\pi\sqrt{LC}}$$

$$= \frac{1}{2\pi\sqrt{(100\ mH)(470\ pF)}}$$

$$= 23{,}226.90\ Hz$$

This frequency of 49.51 kHz is valid because we know that class C amplifiers are only used at RF, or radio frequencies, which are greater than 20 kHz.

The resonant frequency of Example 4 is also a valid RF because it's greater than 20 kHz.

Because the class C amplifier is seldom on, its efficiency is excellent, nearly approaching 100%.

The class C amplifier can provide a large amount of power to its load, so it's often called a large signal amplifier, or power amplifier. For this reason, the class C amplifier is frequently used as the output stage for radio transmitters, which can generate kilowatts of power. Let's build a live version of the class C amplifier.

LAB ACTIVITY 6-2

Class C Amplifier

Materials, Equipment, and Parts:

- *NTE* catalog or Internet access, www.nteinc. com (*Look it up to hook it up.*)

- PC w/Multisim®, Electronics Workbench®, or SPICE.

- DMM with test leads.

- Dual trace oscilloscope w/BNC-to-alligator leads.

- 12 V DC voltage source.

- Function generator, 1 Hz to 30 kHz.

- Breadboard and connecting wires.

- 2N3904 or equivalent NPN transistor.

- 680 pF capacitor; 1 µF electrolytic capacitor (2).

- 47 mH inductor coil.

- 1 kΩ fixed resistor and 100 kΩ fixed resistor.

Discussion Summary:

A class C amplifier uses one transistor in a common-emitter configuration to provide midrange voltage and current gains and a high power gain. Since the class C amplifier is rarely on, its efficiency approaches 100%. However, the class C amplifier misses most of the incoming AC input signal, and the output waveform is a very distorted version of the input waveform. A tank section attached to the collector of the class C amplifier restores the waveform.

Procedure:

SAFETY FIRST. Eye protection should always be worn when working with live voltages. Before powering on a live circuit, always check with your instructor.

1 Write the part number of the transistor at your workstation. Then use the *NTE* catalog or visit the website www.nteinc.com to find and record the NTE replacement number and the diagram number.

Part number _____

NTE number _____

Diagram # _____

2 Draw the transistor outline and label the emitter, base, and collector for the transistor.

3 Build the circuit of Figure 6-8 on your breadboard. This is a class C amplifier without a tank section.

(continues)

LAB ACTIVITY 6-2

(continued)

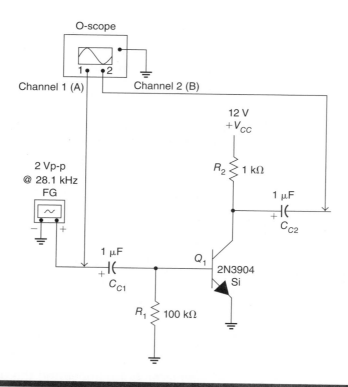

FIGURE 6-8   Class C amplifier without tank section (© Cengage Learning 2012)

4 Connect the positive lead of the 12 V DC voltage source to the top end of resistor R₂.

5 Connect the positive lead of the function generator to the positive side of capacitor C$_{C1}$.

6 Connect the positive lead of the oscilloscope's Channel 1 to the positive side of capacitor C$_{C1}$.

7 Connect the positive lead of the oscilloscope's Channel 2 to the negative side of capacitor C$_{C2}$.

8 Connect the black lead of the function generator, the black lead of the 12 V DC voltage source, the black leads of the o-scope, a wire from the bottom of resistor R₁, and a wire from the emitter of transistor Q₁ to one point (hole) on the breadboard, which is ground.

9 Set the function generator to the sine wave setting and to an amplitude of 2 V peak-to-peak. You may have to use your o-scope's Channel 1 volts/division setting to adjust the voltage to 2 V.

10 Set the function generator for a frequency of 28.1 kHz.

11 Have the instructor check your circuit.

12 Power on your 12 V DC voltage source, the function generator, and the o-scope.

13 Set your o-scope mode switch to dual. Adjust your oscilloscope volts/division settings and time/division settings so that the AC input waveform is on the top of the o-scope display and the AC output waveform is directly below it. What do you notice about the AC output waveform's positive and negative alternations? Describe them below.

LAB ACTIVITY 6-2

14 Measure and record the AC input voltage that appears on the o-scope's Channel 1. You may have to adjust the Channel 1 volts/division setting and/or other o-scope settings. Record the value below.

V_{IN} _____

15 Measure and record the AC output voltage that appears on the o-scope's Channel 2. You may have to adjust the Channel 2 volts/division setting and/or other o-scope settings. Record the value below.

V_{OUT} _____

16 Use the equation below to calculate and record the dynamic voltage gain of the class C amplifier.

$$A_V = \frac{V_{OUT}}{V_{IN}} =$$

A_V _____

17 Power off the 12 V DC source, the function generator, and the o-scope. Remove resistor R_2 (the 1 kΩ resistor) from the circuit. Connect the tank section (capacitor C_3 and coil L_1) to the collector of Q_1 and connect the 12 V V_{CC} to the top end of the tank section as shown in Figure 6-9. Now, we have a class C amplifier with a tank section.

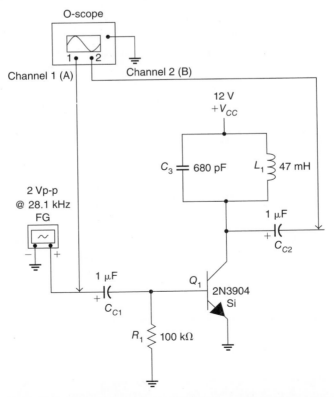

FIGURE 6-9 Class C amplifier with tank section (© Cengage Learning 2012)

(continues)

LAB ACTIVITY 6-2

(continued)

18 Use the resonant frequency equation to calculate the resonant frequency of the tank section. Record the frequency below.

$$f_r = \frac{1}{2\pi\sqrt{LC}} =$$

f_r _____

19 Power on the 12 V DC source, the function generator, and the o-scope. Has the AC output waveform changed? Did the tank circuit restore the AC output waveform to a sine wave?

20 Set your DMM to measure frequency. Connect the positive lead of the DMM to the negative side of capacitor C_{C2}. Connect the negative lead of your DMM to ground. Power on your DMM and measure and record the frequency.

f_r _____

How does the measured resonant frequency of Step 20 compare to the calculated resonant frequency of Step 18? Explain any differences.

21 Measure and record the AC input voltage that appears on the o-scope's Channel 1. It should still be about 2 V_{P-P}. Record the value below.

V_{IN} _____

22 Measure and record the AC output voltage that appears on the o-scope's Channel 2. You may have to adjust the Channel 2 volts/division setting and/or other o-scope settings. Record the value below.

V_{OUT} _____

23 Use the equation below to calculate and record the dynamic voltage gain of the class C amplifier.

$$A_V = \frac{V_{OUT}}{V_{IN}} =$$

A_V _____

Compare this new voltage gain to the dynamic voltage gain measured in Step 16. Did the addition of the tank section increase or decrease the voltage gain?

24 Have the instructor verify your results.

25 Build the circuit on Multisim®, Electronics Workbench®, or SPICE and then compare the results.

CHAPTER SUMMARY

In addition to having a configuration, every transistor amplifier belongs to a particular class. An amplifier class is determined by the way an amplifier is biased; that is, the various DC voltages that affect the transistor's operation. The three transistor operating conditions are cutoff, midpoint, and saturation. At cutoff, a transistor is totally off, I_C is equal to 0 A, and V_{CE} is equal to maximum (V_{CC}). At midpoint biasing, a transistor's I_C is about one-half of maximum, and its V_{CE} is about one-half of maximum. At saturation, a transistor is totally on, I_C is equal to maximum, and $V_{CE} = 0$ V. Driving a transistor amplifier beyond saturation can result in clipping of the waveform.

Each amplifier class—A, AB, B, and C—has unique values of fidelity and efficiency. The fidelity of an amplifier is its ability to accurately reproduce an AC input signal. Distortion is an unwanted change in the shape of an AC signal. High fidelity means little or no distortion. An AC signal entering an amplifier steals power from the V_{CC} (DC supply voltage) to get larger. The efficiency of an amplifier is the percentage of DC power that actually makes it to the amplifier's load, such as another amplifier stage or a speaker.

The class A amplifier is biased at midpoint and has very high fidelity and an efficiency of up to 25%. Class A amplifiers usually use voltage-divider bias, and the V_{CE} must be calculated or measured to verify the amplifier's midpoint biasing. Class A amplifiers are used as the input stage for audio (signals between 20 Hz and 20 kHz) amplifiers.

Class B amplifiers are biased at cutoff and have high fidelity and an efficiency of up to 78.5%. A class B amplifier uses two transistors, one to reproduce the positive alternation of the AC input signal and the other to reproduce the negative alternation. Because of the slow turn-on and turn-off times, a class B amplifier has crossover distortion. The class AB amplifier eliminates crossover distortion by using two compensating diodes at the inputs of the class B transistors. Class B and class AB amplifiers are used in the intermediate or output stages of audio amplifiers, and they can be used to drive speakers. Many class B and AB amplifiers are manufactured as ICs.

The class C amplifier is biased way below cutoff and has low fidelity and an efficiency of up to 99%. Since class C amplifiers are on for less than one-half of the time, they need a tank section connected to the transistor's collector terminal to restore the AC waveform. The tank section provides a frequency to match the frequency of the AC input signal. Class C amplifiers supply a great deal of power to the load and are often used as the output stage for radio receivers and transmitters. Radio frequencies are greater than 20 kHz.

CHAPTER EQUATIONS

Amplifier efficiency

(Equation 6-1)

$$\eta = \frac{P_{AC}}{P_{DC}}(100)$$

η = % efficiency

P_{AC} = AC output signal power

P_{DC} = DC input power (V_{CC})

Class C amplifier resonant frequency of tank section

(Equation 6-2)

$$f_r = \frac{1}{2\pi\sqrt{LC}}$$

f_r = resonant frequency in Hz

L = inductance of coil in tank

C = capacitance of capacitor in tank

CHAPTER REVIEW QUESTIONS

Chapter 6-1

1. If an amplifier has high distortion, what type of fidelity will it have?

2. An amplifier has an AC output power of 600 mW and a DC input power of 2.5 W. What is the efficiency of the amplifier?

3. An amplifier has an AC output power of 270 W and a DC input power of 301 W. What is the efficiency of the amplifier?

Chapter 6-2

4. Do class A amplifiers have high efficiency or high fidelity?

5. What is the collector current of an amplifier during cutoff?

6. What is the collector current of an amplifier during saturation?

7. What is the collector current of an amplifier biased at midpoint?

8. How many watts of power can a small signal amplifier safely handle?

Chapter 6-3

9. Which amplifier class is shown in Figure 6-10?

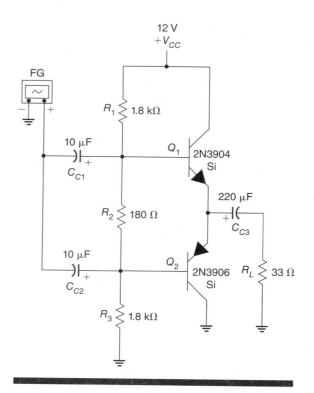

FIGURE 6-10   Amplifier class (© Cengage Learning 2012)

10. What is the key visual feature of a class B amplifier?

11. What is the typical efficiency of a class B amplifier?

12. Explain push-pull operation in a class B amplifier.

13. Draw a waveform that shows the effects of crossover distortion.

Chapter 6-4

14. How do compensating diodes eliminate crossover distortion in an amplifier?

15. What is the typical output power of a large signal amplifier?

16. Name one application for a class AB amplifier.

17. Which amplifier class is shown in Figure 6-11?

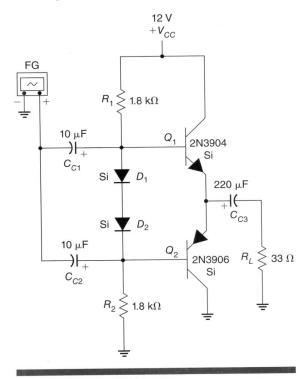

FIGURE 6-11 Amplifier class (© Cengage Learning 2012)

Chapter 6-5

18. Which amplifier class is biased below cutoff?

19. What is the key visual feature of a class C amplifier?

20. Which amplifier class would most likely be used at the output of a 100 kW radio transmitter?

21. What is the resonant frequency of a class C amplifier whose tank section has a coil measuring 33 mH and a capacitor measuring 220 pF?

22. What is the resonant frequency of the tank section for the class C amplifier in Figure 6-12?

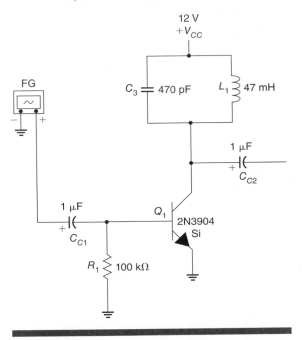

FIGURE 6-12 Class C amplifier (© Cengage Learning 2012)

Chapter 6-1 through 6-5

23. Which amplifier class would most likely be found at the input stage of a radio receiver?

24. The V_{CC} of an amplifier is 14 V. The measured V_{CE} is 4 V. Is the amplifier biased for class A operation?

25. Which amplifier class uses a common-collector configuration?

26. What is the approximate V_{CE} of a transistor at saturation?

27. Which amplifier class uses two transistors and is subject to crossover distortion?

28. Which amplifier class has the highest fidelity?

29. Which amplifier class has a tank section connected to its collector to restore the AC input signal?

30. Which amplifier class has compensating diodes to eliminate crossover distortion?

Field Effect Transistors (FETs)

OBJECTIVES   *Upon completion of this chapter, you should be able to:*

- Draw and label the schematic symbol for an N-channel JFET.

- Explain the operation of an N-channel JFET.

- Distinguish between pinch-off voltage (V_P) and $V_{GS\ (off)}$ in a JFET.

- Use a semiconductor replacement guide to determine critical information about JFETs and MOSFETs.

- Identify an FET's drain, gate, and source leads.

- Use a DMM to test a JFET and MOSFET.

- Build a circuit using a JFET as a high-speed switch.

- List the distinctive features of the common-source (CS), common-drain (CD), and common-gate (CG) amplifier configurations.

- Calculate the transconductance of a JFET.

- Calculate the static voltage gain of a CS amplifier.

- Build a CS amplifier circuit to calculate and measure its dynamic voltage gain.

- Distinguish between depletion-mode operation and enhancement-mode operation.

- List two methods to reduce the effects of ESD.

- Identify the schematic symbols for a D-MOSFET, an E-MOSFET, and an IGBT.

MATERIALS, EQUIPMENT, AND PARTS

Materials, equipment, and parts needed for the lab experiments in this chapter are listed below:

- *NTE* catalog or Internet access, www.nteinc.com
- PC w/Multisim®, Electronics Workbench®, or SPICE.
- Grounded wrist strap or anti-static mat.
- DMM with test leads.
- Dual trace oscilloscope w/BNC-to-alligator leads.
- 12 V DC voltage source.
- 15 V DC voltage source.
- Function generator, 1 Hz to 10 kHz.
- Breadboard and connecting wires.

- 2N5485 or equivalent N-channel JFET.
- One N-channel JFET such as 2N5485, 2N5457, or 2N5459.
- Two N-channel E-MOSFETs such as BS170, 2N7000, IRF510, IRF620, or equivalent.
- One NPN BJT such as 2N3904, 2N2219, or 2N3053.
- 2N3904 or equivalent NPN transistor.
- 2N3906 or equivalent PNP transistor.
- Two-terminal LED (any color).

- 1 µF electrolytic capacitor (2), 10 µF electrolytic capacitor (2), 22 µF electrolytic capacitor, and 220 µF electrolytic capacitor.
- 47 mH inductor coil.
- 33 Ω fixed resistor, 220 Ω fixed resistor, 680 Ω fixed resistor, 1 kΩ fixed resistor, 1.2 kΩ fixed resistor, 1.5 kΩ fixed resistor, 2.2 kΩ fixed resistor (2), 2.7 kΩ fixed resistor, 4.3 kΩ fixed resistor, 100 kΩ fixed resistor, 1 MΩ fixed resistor, and 1.2 MΩ fixed resistor.

GLOSSARY OF TERMS

Field effect transistor (FET) A three-terminal voltage-controlled semiconductor device used as a high-speed switch or as an amplifier

N-channel JFET A junction field effect transistor that controls the movement of electrons through a channel of N-type material

P-channel JFET A junction field effect transistor that controls the movement of holes through a channel of P-type material

Pinch-off voltage, or V_P The operating condition of a field effect transistor where increasing the V_{DS} causes no further increase in I_D; it is similar to saturation in a BJT

Gate-to-source cutoff voltage, or $V_{GS\ (off)}$ The operating condition of a field effect transistor where the reverse bias and depletion region are at a maximum, so no drain current or source current flows; it is similar to cutoff in a BJT

V_{DD} The main DC supply voltage for an FET amplifier, which is a positive voltage applied across the entire transistor with respect to ground

Common-source amplifier The most popular JFET amplifier configuration, which has the input signal applied to the gate terminal and the output signal appearing at the drain terminal

Common-drain amplifier The JFET amplifier configuration that has the input signal applied to the gate and the output signal appearing at the source

Common-gate amplifier The JFET amplifier configuration that

has the input signal applied to the source and the output signal appearing at the drain

Transconductance The amplification rating for a field effect transistor, which is the result of a change in I_D divided by a change in V_{GS}; it shows how a change in V_{GS} causes a change in I_D

MOSFET (Metal Oxide Semiconductor Field Effect Transistor) A type of insulated gate field effect transistor used as a high-speed switch or as an amplifier

IGFET A category of field effect transistors with extremely high input impedance, characterized by a gate terminal made of an insulating material such as silicon dioxide or hafnium dioxide

Electrostatic discharge or ESD The release of static electricity that can destroy the gate material in IGFETs; it is often a silent killer, more intense during cool, dry weather

D-MOSFET A metal oxide semiconductor field effect transistor that can be operated in depletion mode or enhancement mode

E-MOSFET A metal oxide semiconductor field effect transistor that can be operated only in enhancement mode

Depletion mode A method of operation for a JFET or MOSFET in which the transistor is totally on (maximum I_D) when the V_{GS} is 0 V, and by making V_{GS} more negative, the flow of electrons is reduced or depleted

Enhancement mode A method of operation for a MOSFET where a transistor has a positive V_{GS} that forms a "bridge" between the drain and source, causing current to flow through the transistor

Threshold voltage The voltage level at which an E-MOSFET forms a bridge across its channel and starts conducting current

MuGFET (Multiple Gate Field Effect Transistor) A type of field effect transistor constructed of two or more gates that reduce the surface area and thus the overall input capacitance of the transistor

IGBT A type of three-terminal insulated gate transistor that is a hybrid of a BJT and MOSFET in terms of construction and operation; it is a fast-switching transistor capable of handling large current and power requirements

7-1 INTRODUCTION TO FETs

Six years after the Bipolar Junction Transistor (BJT) was invented in 1947 by John Bardeen, Walter Brattain, and William Shockley at Bell Labs in Murray Hill, New Jersey, the Field Effect Transistor (FET) was invented by G. C. Dacey and Ian Munro Ross at the same Bell Labs. Although the BJT became the first popular transistor in industrial applications, the FET would later prove to have more advantages over its BJT counterpart. BJTs and FETs are the two major categories of transistors. Since we're already familiar with the BJT, throughout this chapter we'll compare and contrast these two types of transistors.

Like the BJT, the **Field Effect Transistor** is a three-terminal semiconductor device that is used as a high-speed switch or as an amplifier. BJTs and FETs differ in internal construction and method of operation. BJTs are current-controlled devices whose operation depends on the interaction of electrons *and* holes; that's why they are called bipolar. FETs are voltage-controlled devices whose operation depends on the movement of either electrons *or* holes; that's why they are called unipolar. FETs look like BJTs, so the only way to tell the difference is *Look it up to hook it up.*

The two major categories of FETs are Junction Field Effect Transistors (JFETs, or "JAY-FETS") and Insulated Gate Field Effect Transistors (IGFETs). The most common IGFET is the Metal Oxide Semiconductor Field Effect Transistor (MOSFET, or "MOSS-FET"). JFETs and MOSFETs differ in construction, but they are both voltage-controlled devices. In this chapter, we'll discuss the construction, operation, and testing of both JFETs and MOSFETs.

7-2 JUNCTION FIELD EFFECT TRANSISTOR (JFET) BASICS

JFETs are constructed in one of two ways: N-channel or P-channel. Figure 7-1 shows simplified drawings of both FET constructions and their schematic symbols.

Figure 7-1(a) shows the **N-channel JFET**, the type used in most electronic circuits. Notice that

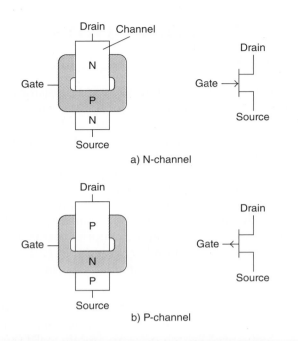

FIGURE 7-1   JFET constructions and schematic symbols: a) N-channel and b) P-channel (© Cengage Learning 2012)

the N-channel JFET has one part N-type material, which is usually arsenic with excess electrons, and one part P-type material, which is usually boron with excess holes (lacking electrons); thus, the JFET more closely resembles a diode in construction. (The NPN transistor discussed in Chapter 4 had *two* parts N-type material and *one* part P-type material.) This is one of the major differences between a JFET and a BJT.

It may not be clear from Figure 7-1(a), but the P-type material actually wraps around and blends with the N-type material, much in the way a label does on a can of soup. Because of this arrangement, there is only one PN junction in a JFET. Also, you can't split a JFET into separate P-type and N-type materials because these materials are chemically combined with the silicon base. The same holds true for PN junction diodes and BJTs.

Like the BJT, the JFET has three terminals: drain, gate, and source (from top to bottom), so I tell my students to think of Dollar General Store to help them remember the terminals. For the N-channel JFET of Figure 7-1(a), the drain terminal and the source terminal are connected to the same N-type material that has a large number of excess electrons. The gate terminal is connected to P-type material that has a small number of excess holes. The channel connecting the drain to the

source is made of N-type material, so that's why it's called an N-channel JFET.

For the **P-channel JFET** of Figure 7-1(b), the drain and source terminals are connected to the same P-type material. The gate is connected to the N-type material. The channel connecting the drain to the source is made of P-type material, so that's why it's called a P-channel JFET.

Notice in both schematic symbols of Figure 7-1 that the gate terminal, or lead, has an arrow, which is used to identify the type of JFET—N-channel or P-channel. If the arrow of the gate is pointing i**N** as shown in Figure 7-1(a), it's an N-channel JFET. If the arrow of the gate is pointing out as shown in Figure 7-1(b), then it is a P-channel JFET. A technician must be able to identify the type of JFET and each of the leads to test it correctly.

Like BJTs, JFET part numbers often have the prefix 2N, and each JFET has a manufacturer specification (spec) sheet or sheets, which list everything about the JFET. Figure 7-2 shows two spec sheets for a 2N5485 JFET. (Actually, the spec sheets list the characteristics for several JFETs; sometimes semiconductor manufacturers will group similar JFETs on one or more spec sheets.) The spec sheets show the JFET's applications; maximum ratings; thermal characteristics; and electrical characteristics—anything you want to know about the JFET. Figure 7-2 shows only the first two of seven sheets of information about the 5485 JFET.

If you don't have access to spec sheets, then you can use a semiconductor replacement guide such as the *NTE* catalog to look up a JFET. You should be pretty sure by now in using the *NTE* catalog. Remember, the procedure is the same for looking up JFETs: start at the back of the book, go to the front of the book, and then go to the middle of the book. Back, front, middle. Keep in mind that the main reason we use a semiconductor replacement guide is to identify the source, gate, and drain terminals of a JFET so that we can test it or connect it properly in a circuit.

Looking again at Figure 7-1, you can see that both the N-channel and P-channel JFETs have only one junction or depletion region created during the manufacturing process. In the N-channel JFET, this occurs where the wraparound P-type material meets the N-type material. For the P-channel JFET, this junction or depletion region exists where the wraparound N-type material meets the P-type material. This fact will affect the way we test

2N5484/5485/5486 MMBF5484/5485/5486 —

February 2009

2N5484/5485/5486 MMBF5484/5485/5486

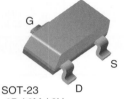

G
S D   TO-92

G
S
D
SOT-23
Mark: 6B / 6M / 6H

NOTE: Source & Drain
are interchangeable

N-Channel RF Amplifier

This device is designed primarily for electronic switching
applications such as low On Resistance analog switching
Sourced from Process 50.

Absolute Maximum Ratings*   TA = 25°C unless otherwise noted

Symbol	Parameter	Value	Units
V_{DG}	Drain-Gate Voltage	25	V
V_{GS}	Gate-Source Voltage	−25	V
I_{GF}	Forward Gate Current	10	mA
T_J, T_{stg}	Operating and Storage Junction Temperature Range	−55 to +150	°C

*These ratings are limiting values above which the serviceability of any semiconductor device may be impaired.

NOTES:
1) These ratings are based on a maximum junction temperature of 150 degrees C.
2) These are steady state limits. The factory should be consulted on applications involving pulsed or low duty cycle operations.

Thermal Characteristics   TA = 25°C unless otherwise noted

Symbol	Characteristic	Max 2N5484–5486	Max *MMBF5484–5486	Units
P_D	Total Device Dissipation Derate above 25°C	350 2.8	225 1.8	mW mW/°C
$R_{\theta JC}$	Thermal Resistance, Junction to Case	125		°C/W
$R_{\theta JA}$	Thermal Resistance, Junction to Ambient	357	556	°C/W

*Device mounted on FR-4 PCB 1.6" × 1.6" × 0.06".

www.fairchildsemi.com

FIGURE 7-2(a)   Specification sheets for a 2N5485 N-channel JFET (Courtesy of Fairchild Semiconductor®)

N-Channel RF Amplifier
(continued)

Electrical Characteristics   TA = 25°C unless otherwise noted

Symbol	Parameter	Test Conditions	Min	Typ	Max	Units
OFF CHARACTERISTICS						
$V_{(BR)GSS}$	Gate-Source Breakdown Voltage	$I_G = -1.0 \ \mu A$, $V_{DS} = 0$	−25			V
I_{GSS}	Gate Reverse Current	$V_{GS} = -20$ V, $V_{DS} = 0$ $V_{GS} = -20$ V, $V_{DS} = 0$, $T_A = 100°C$			−1.0 −0.2	nA μA
$V_{GS(off)}$	Gate-Source Cutoff Voltage	$V_{DS} = 15$ V, $I_D = 10$ nA **5484** **5485** **5486**	−0.3 −0.5 −2.0		−3.0 −4.0 −6.0	V V V
ON CHARACTERISTICS						
V_{DSS}	Zero-Gate Voltage Drain Current*	$V_{DS} = 15$ V, $V_{GS} = 0$ **5484** **5485** **5486**	1.0 4.0 8.0		5.0 10 20	mA mA mA
SMALL SIGNAL CHARACTERISTICS						
g_{fs}	Forward Transfer Conductance	$V_{DS} = 15$ V, $V_{GS} = 0$, $f = 1.0$ kHz **5484** **5485** **5486**	3000 3500 4000		6000 7000 8000	μmhos μmhos μmhos
$Re_{(Yis)}$	Input Conductance	$V_{DS} = 15$ V, $V_{GS} = 0$, $f = 100$ MHz **5484** $V_{DS} = 15$ V, $V_{GS} = 0$, $f = 400$ MHz **5485/5486**			100 1000	μmhos μmhos
g_{os}	Output Conductance	$V_{DS} = 15$ V, $V_{GS} = 0$, $f = 1.0$ kHz **5484** **5485** **5486**			50 60 75	μmhos μmhos μmhos
$Re_{(Yos)}$	Output Conductance	$V_{DS} = 15$ V, $V_{GS} = 0$, $f = 100$ MHz **5484** $V_{DS} = 15$ V, $V_{GS} = 0$, $f = 400$ MHz **5485/5486**			75 100	μmhos μmhos
$Re_{(Yfs)}$	Forward Transconductance	$V_{DS} = 15$ V, $V_{GS} = 0$, $f = 100$ MHz **5484** $V_{DS} = 15$ V, $V_{GS} = 0$, $f = 400$ MHz **5485** **5486**	2500 3000 3500			μmhos μmhos μmhos
C_{iss}	Input Capacitance	$V_{DS} = 15$ V, $V_{GS} = 0$, $f = 1.0$ MHz			5.0	pF
C_{rss}	Reverse Transfer Capacitance	$V_{DS} = 15$ V, $V_{GS} = 0$, $f = 1.0$ MHz			1.0	pF
C_{oss}	Output Capacitance	$V_{DS} = 15$ V, $V_{GS} = 0$, $f = 1.0$ MHz			2.0	pF
NF	Noise Figure	$V_{DS} = 15$ V, $R_G = 1.0$ kΩ, $f = 100$ MHz **5484** $V_{DS} = 15$ V, $R_G = 1.0$ kΩ, $f = 400$ MHz **5484** $V_{DS} = 15$ V, $R_G = 1.0$ kΩ, $f = 100$ MHz **5485/5486** $V_{DS} = 15$ V, $R_G = 1.0$ kΩ, $f = 400$ MHz **5485/5486**		 4.0	3.0 2.0 4.0	dB dB dB dB

www.fairchildsemi.com

FIGURE 7-2(b)   (continued)

an N-channel JFET. For testing purposes, an N-channel JFET is really just one PN junction "diode," but this diode *can be measured from either the gate lead to the drain lead or from the gate lead to the source lead* because both the drain and source leads are connected to the same N-type material. Figure 7-3 shows these two "diodes" (the blue triangles): one between the gate lead and the drain lead in Figure 7-3(a), and the other between the gate lead and the source lead in Figure 7-3(b).

Remember that a force, or barrier voltage, exists between P-type material and N-type material, @ 0.3 V for germanium and @ 0.6 V for silicon. When you apply a positive voltage to the P-type material and a negative voltage to the N-type material, this forward biases the JFET (positive to positive and negative to negative), the barrier or depletion region breaks down, and current flows through the JFET. Since the JFET is a voltage-controlled device, there really should be *no* current through the gate. *A gate current greater than 50 mA might actually destroy a JFET.* However, using the diode setting of a DMM to forward bias the gate-to-source junction only produces a few mA through the gate, so testing a JFET with the diode setting is harmless.

On the other hand, when you apply a positive voltage to the N-type material and a negative voltage to the P-type material, this reverse biases the JFET (positive to negative and negative to positive), the depletion region increases, and no current flows through the JFET.

Like the BJT, we can test a JFET using the diode setting of a digital multimeter. Remember, when you check a JFET (or BJT) on the diode setting, the numerical result is in volts, though the meter may *not* show "volts" or the letter *V*. Testing a JFET is similar to testing a BJT except that the results will be somewhat different. Use the following steps for testing an N-channel JFET.

Testing a JFET on the diode setting requires six checks. First, set your meter's rotary selector switch to the diode setting. Then touch or clip the red (positive) lead of the meter to the source terminal of the JFET, and touch or clip the black (negative) lead of the meter to the gate terminal of the JFET. This reverse biases the PN junction diode between the gate and source, so the meter should display infinite, which may be represented by the infinity symbol (∞) or "OL" (overload) for a good N-channel JFET.

For the second check, touch or clip the red (positive) lead of the meter to the source of the JFET, and touch or clip the black (negative) lead of the meter to the drain of the JFET. Since the source and drain are actually connected to the same N-type material (the channel), the measurement should be a low voltage drop (@ 0.3 V) for a good N-channel JFET. The meter will probably fluctuate and then settle on @ 0.3 V.

For the third check, touch or clip the red (positive) lead of the meter to the gate of the JFET, and touch or clip the black (negative) lead of the meter to the source of the JFET. This forward biases the PN junction between the gate and source, current flows through the depletion region, and the meter should display between 0.6 to 0.7 (volts) for a good N-channel JFET as shown in Figure 7-3(b).

For the fourth check, touch or clip the red (positive) lead of the meter to the gate of the JFET, and touch or clip the black (negative) lead of the meter to the drain of the JFET. This forward biases the same PN junction but now between the gate and the drain, current flows through the depletion region, and the meter should display between 0.6 to 0.7 (volts) for a good N-channel JFET as shown in Figure 7-3(a).

For the fifth check, touch or clip the red (positive) lead of the meter to the drain of the JFET, and touch or clip the black (negative) lead of the meter to the source of the JFET. Since the source and

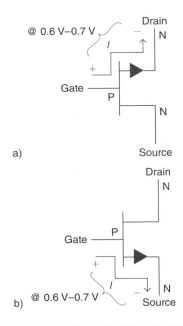

FIGURE 7-3 Forward biasing the depletion region in an N-channel JFET: a) Gate-drain forward biased and b) Gate-source forward biased (© Cengage Learning 2012)

drain are actually connected to the same N-type material (the channel), the measurement should be a low voltage drop (@ 0.3 V) for a good N-channel JFET. The meter will probably fluctuate and then settle on @ 0.3 V.

For the sixth and final check, touch or clip the red (positive) lead of the meter to the drain of the JFET, and touch or clip the black (negative) lead of the meter to the gate of the JFET. Again, this reverse biases the PN junction diode between the drain and the gate so the meter should display infinite, which may be represented by the infinity symbol (∞) or "OL" (overload) for an N-channel JFET.

So, JFET testing using the diode and meter leads requires six checks. The results of testing an N-channel JFET are summarized below:

1) Positive on source and negative on gate—display should be infinite or OL.

2) Positive on source and negative on drain—display should be @ 0.3 (volts).

3) Positive on gate and negative on source—display should be @ 0.6 to 0.7 (volts).

4) Positive on gate and negative on drain—display should be @ 0.6 to 0.7 (volts).

5) Positive on drain and negative on source—display should be @ 0.3 (volts).

6) Positive on drain and negative on gate—display should be infinite or OL.

Notice that there are four numerical readings and two infinite readings for a good JFET. If the ratio is different—three and three or five and one or six and zero—then the JFET is defective and should be replaced. This diode-setting/six-check method provides results that are just the opposite of a BJT: a good BJT has two numerical readings and four infinite readings.

Unfortunately, you can't test JFETs using the h_{FE} setting/socket option available on many DMMs. This is only for BJTs because it measures their DC current gain—I_C divided by I_B. JFETs are voltage-controlled devices, and they technically have no gate current, so they can't have any DC current gain. Besides the diode-setting/six-check method, the only other option for testing a JFET is to use a specialized semiconductor tester such as a curve tracer.

Now that we know the methods for identifying and testing a JFET, it's time to put theory into practice.

LAB ACTIVITY 7-1

Identifying and Testing Junction Field Effect Transistors (JFETs) and Bipolar Junction Transistors (BJTs)

Materials, Equipment, and Parts:

- *NTE Semiconductor Technical Guide and Cross Reference* catalog

  or

  NTE QUICKCross™ software

  or

  Internet access, www.nteinc.com

- DMM with test leads.

- One N-channel JFET such as 2N5485, 2N5457, or 2N5459.

- One NPN BJT such as 2N3904, 2N2219, or 2N3053.

Discussion Summary:

Both Junction Field Effect Transistors (JFETs) and Bipolar Junction Transistors (BJTs) can go bad (become shorted or open). Technicians need to be able to identify the polarity of an FET (JFET or MOSFET) and BJT (NPN or PNP); the material it's made of (N-channel or P-channel) and (silicon or germanium); and the leads (source, gate, drain for JFET and emitter, base, collector for BJT) to test the transistor correctly. In this lab, you will use the diode-setting/six-check method for testing both a JFET and a BJT.

Procedure:

1 For each transistor at your workstation, use the *NTE* catalog to research and record the following information for each of the three transistors. For the first transistor, Steps 1 and 3 have been completed to serve as an example. Transistor 2 must be a BJT, and transistor 3 must be a JFET:

	1	2	3
Part #	2N5458	_____	_____
NTE #	457	_____	_____
Polarity	JFET	_____	_____
Material	N-channel	_____	_____
Description & Application	General-purpose amplifier, switch	_____	_____
Case Style	TO92	_____	_____
Diagram #	9e	_____	_____

(continues)

LAB ACTIVITY 7 - 1

(continued)

2   Go to the diagram page for each transistor and then draw the transistor outline (TO) for each transistor. Label the emitter, base, and collector terminals for the BJT and the source, gate, and drain terminals for the FET.

1   **2**   **3**

3   Use your DMM's diode setting and meter leads to check each transistor. Record the results. For the first volt reading, the positive lead (+) of the DMM is connected to the source (S) and the negative lead (−) of the DMM is connected to the gate terminal (G). Perform the remaining checks in this manner. Note that transistors 1 and 3 are JFETs.

	1			**2**			**3**
+ −	Volt reading	+ −	Volt reading	+ −	Volt reading
S G	OL	E B	_____	S G	_____
S D	0.289 V	E C	_____	S D	_____
G S	0.736 V	B E	_____	G S	_____
G D	0.734 V	B C	_____	G D	_____
D S	0.302 V	C E	_____	D S	_____
D G	OL	C B	_____	D G	_____
Good?	Yes		_____		_____

4   Out of six checks, how many infinite readings should a good JFET measure? Out of six checks, how many infinite readings should a good BJT measure?

5   Have the instructor check your results.

7-3 JFET BIASING AND OPERATION

Now that we understand how to identify and test a JFET, it's time to investigate exactly how a JFET works. The JFET is a voltage-controlled device, so its behavior will be quite different from that of a BJT. Our discussion will focus on the most widely used JFET, the N-channel JFET.

A DC voltage source, or bias, causes current to flow in a JFET. Like the BJT, the N-channel JFET needs two DC voltages (biases) to operate: a reverse bias on the gate-to-source junction, called the V_{GS}, and a reverse bias on the drain-to-source junction, called the V_{DS}. Figure 7-4 shows three examples of a properly biased N-channel JFET, NTE replacement number 133. These circuits are for learning purposes only.

If you look carefully at Figure 7-4(a), you will see that the gate-to-source junction is reverse biased. For the JFET, the gate terminal acts like a valve, controlling the voltage and thus the current flow through the JFET. The voltage applied to the gate and source, or V_{GS}, must be either 0 volts or a negative voltage. To accomplish this, the gate-to-source junction must be reverse biased. The negative end of the 0 V DC voltage source on the left is connected to the gate terminal (made of P-type material), and the positive end of the 0 V DC voltage source is connected to the source terminal, which in turn is connected to the N-channel. Thus, we have negative to positive and positive to negative—reverse bias. This DC voltage on the left controls the gate-to-source voltage, or V_{GS}.

On the right side of the JFET, the positive end of the 6 V DC voltage source is connected to the drain terminal, which in turn is connected to the N-channel so that the drain is reverse biased. This 6 V DC voltage source on the right controls the drain-to-source voltage, or V_{DS}.

How exactly does the circuit of Figure 7-4 operate? The 6 V DC voltage source (V_{DS}) on the right causes conventional, or hole, current to flow down through the N-channel. If the DC voltage source on the left that controls the V_{GS} is set to 0 V as shown in Figure 7-4(a), then the drain current (I_D) is maximum, which for this JFET (NTE 133) is 14.6 mA. This value is consistent with the "Drain Current Max (OFF)" value listed in the

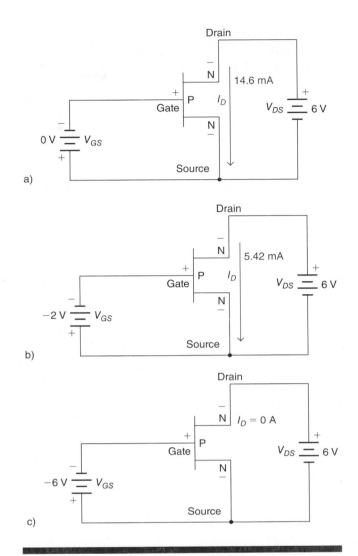

a)

b)

c)

FIGURE 7-4 N-channel JFET with gate-to-source and drain-to-source biasing: a) V_{GS} @ 0 V, b) V_{GS} @ –2 V, and c) V_{GS} @ –6 V (© Cengage Learning 2012)

NTE manual. A spec sheet often lists the maximum drain current value under the heading I_{DSS}.

Thus, the JFET is totally *on* when the gate voltage is 0 V; in contrast, the BJT is totally *off* when its base voltage is 0 V. The source current in the JFET is also at maximum because both the drain and source currents travel in the same N-channel; thus, $I_D = I_S$. Current is maximum because the depletion region is small where the gate and its wraparound P-type material meet the N-type material of the N-channel. (Electron flow is opposite to hole current flow, so electrons are moving from the source terminal to the "drain" terminal.)

Notice that no current flows into the gate terminal because it is reverse biased. Remember, a gate current greater than 50 mA can destroy a JFET. If you reach a point where increasing the V_{DS} causes no more increase in I_D, the condition is similar to saturation in a BJT, and it is called (oddly) the **pinch-off voltage, or V_P**. In Figure 7-4(a), the V_P is 6 V.

In Figure 7-4(b), the V_{GS} has been changed to -2 V. By making V_{GS} more negative, we've increased the reverse bias. As reverse bias increases, the depletion region between the N-channel and the P-type material attached to the gate terminal also increases. This increased depletion region reduces the drain current—and thus the source current—to 5.42 mA. This reverse bias and increased depletion region chokes the flow of electrons in the N-channel, much like squeezing a can around the middle.

In Figure 7-4(c), the V_{GS} has been changed to -6 V. The reverse bias and depletion region are at maximum, so no drain current or source current flows. This condition is like cutoff in a BJT, and it is called the **gate-to-source cutoff voltage, or $V_{GS \, (off)}$**. Notice that the $V_{GS \, (off)}$ and the V_P in Figure 7-4 are the same value—6 V—but with opposite polarities. This is typically the case for JFETs. The *NTE* manual does not include the negative sign ($-$) for values of $V_{GS \, (off)}$ because $V_{GS \, (off)}$ is understood to be a negative value. $V_{GS \, (off)}$ is the negative value of gate-to-source voltage that reduces current flow in a JFET to 0 A.

In summary, the JFET has the following operating conditions:

V_P, or pinch-off voltage (saturation): totally on, $I_D =$ maximum, $V_{GS} = 0$ V.

Constant-current region: I_D determined by negative values of V_{GS}.

$V_{GS \, (off)}$ (cutoff): $I_D = 0$ A, $V_{GS} = V_P$ (same value but opposite polarity).

In all three cases, V_{GS} *controls* I_D, and the $I_D = I_S$ since they both travel in the same N-channel. Also, in all three cases there is no I_G because the gate terminal is reverse biased. Remember, the gate-source junction is never forward biased; a gate current greater than 50 mA can break down the junction and destroy a JFET.

7-4 THE JFET IN SWITCHING APPLICATIONS

In Chapters 4, 5, and 6, we learned that a transistor is used as a high-speed switch or as an amplifier, and we focused on the BJT. A JFET can also be used as a high-speed switch or as an amplifier. In this section, we'll look at how a JFET can be used to rapidly switch a load on and off. Figure 7-5 shows a JFET in a switching application.

The circuit of Figure 7-5 has a function generator on the left that will supply an AC voltage (sine wave) of 5 V @ 1 Hz to the gate-to-source junction of the 5485 N-channel JFET. The gate resistor R_G and the source resistor R_S work together to reverse bias the gate-to-source junction of the JFET. The 12 V DC voltage source on the right provides drain current. The LED connected to the drain terminal is the load. The 680 Ω resistor R_D limits the current through the LED.

How does the JFET operate as a high-speed switch? The 12 V DC source on the right causes I_D to flow through the 680 Ω drain resistor R_D, through the LED, and through the 1.5 kΩ source resistor R_S to ground. The LED has a positive voltage on its anode and a negative voltage on its cathode, so it lights.

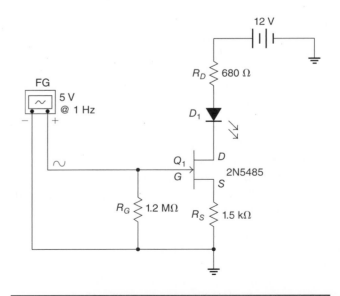

FIGURE 7-5 2N5485 N-channel JFET used as a high-speed switch (© Cengage Learning 2012)

When the AC voltage from the function generator is positive, the LED remains on. When the AC voltage from the function generator goes negative, it reverse biases the gate-to-source junction. At some point, the AC voltage is so *negative* that the reverse bias and depletion region of the JFET are at the maximum (gate-to-source cutoff voltage, or $V_{GS\ (off)}$), and no I_D or I_S flows. Without any I_D, the LED turns off. Thus, the alternating voltage from the function generator turns the LED *off*.

The JFET and BJT behave differently in switching applications. In Lab Activity 4-3, we built a circuit using the BJT as a switch. In that lab, the positive alternation of the AC voltage from the function generator forward biased the BJT, caused I_C to flow, and lit the LED.

For the JFET in Figure 7-5, the *negative* alternation of the AC voltage reverse biases the JFET, stops the flow of I_D, and turns off the LED. Thus, both JFETs and BJTS can be used as high-speed switches. The difference is in the way they operate.

For Figure 7-5, we initially have the function generator set to 1 Hz. If you increase the frequency of the function generator, then the LED will blink rapidly until you reach a frequency where your eyes see the LED as just one steady light.

Let's build it live.

The JFET as a High-Speed Switch

Materials, Equipment, and Parts:

- *NTE* catalog or Internet access, www.nteinc. com (*Look it up to hook it up.*)

- PC w/Multisim®, Electronics Workbench®, or SPICE.

- DMM with test leads.

- 12 V DC voltage source.

- Function generator, 1 Hz to 60 Hz.

- Breadboard and connecting wires.

- 2N5485 or equivalent N-channel JFET.

- Two-terminal LED (any color).

- 680 Ω fixed resistor, 1.5 kΩ fixed resistor, and 1.2 MΩ fixed resistor.

Discussion Summary:

A JFET can be used as a high-speed switch. A 12 V DC voltage source provides I_D to light an LED. When a negative alternating voltage reverse biases the gate-to-source junction of a JFET to the point of gate-to-source cutoff voltage, or $V_{GS \, (off)}$, the I_D stops flowing. If an LED is part of the drain circuit, then the LED will turn off.

Procedure:

SAFETY FIRST. Eye protection should always be worn when working with live voltages. Before powering on a live circuit, always check with your instructor.

1 Write the part number of the JFET at your workstation. Then use the *NTE* catalog or visit the website www.nteinc.com to find and record the NTE replacement number and the diagram number.

Part number _____

NTE number _____

Diagram # _____

2 Draw the JFET transistor outline and label the drain, source, and gate.

(continues)

LAB ACTIVITY 7-2

(continued)

3 Build the circuit of Figure 7-6 on your breadboard.

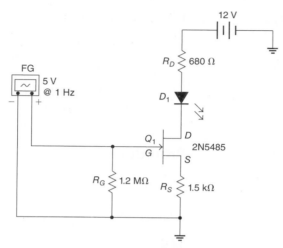

FIGURE 7-6   2N5485 N-channel JFET used as a high-speed switch (© Cengage Learning 2012)

4 Connect the positive lead of the function generator to the gate of JFET Q_1.

5 Connect the positive lead of the 12 V DC voltage source to one end of the 680 Ω resistor R_D.

6 Connect the black lead of the function generator, the black lead of the 12 V DC voltage source, a wire from the bottom of the 1.5 kΩ resistor R_S, and a wire from the bottom of the 1.2 MΩ resistor R_G to one point (hole) on the breadboard, which is ground.

7 Set the function generator to the sine wave setting and to an amplitude of 5 V. You may have to use your DMM set on the AC volts setting to adjust the voltage to 5 V.

8 Set the function generator for a frequency of 1 Hz (or a little higher).

9 Have the instructor check your circuit.

10 Power on your 12 V DC voltage source. Does the LED turn on? Yes _____ No _____

11 Power on your function generator? Does the LED turn off *and* on? Yes _____ No _____

12 Slowly increase the frequency of the function generator until the LED is a steady light. Record the frequency _____

13 Why does the LED become a steady light at the frequency of Step 12?

14 Have the instructor verify your results.

15 Build the circuit on Multisim®, Electronics Workbench®, or SPICE and then compare the results.

7-5 VOLTAGE-DIVIDER BIAS IN JFETs

Up to this point, we've learned how a JFET operates and how to connect it as a high-speed switch. A JFET can also be used as an amplifier. Remember, an amplifier takes a small signal (usually AC) and makes it larger. Like the BJT amplifier, the JFET amplifier has three important characteristics: input impedance (Z_{IN}), output impedance (Z_{OUT}), and gain. However, these characteristics of a JFET amplifier differ from those of a BJT amplifier.

Because a JFET's gate-to-source junction is always reverse biased, it has no I_G, and its gate input impedance is incredibly high—in the $M\Omega$ to $G\Omega$ range. This can be a great advantage in certain applications. In multistage amplifiers, the input impedance of one stage can affect the stage that comes before it, the one feeding it. For example, we know from Chapter 5 that a BJT common-emitter amplifier has a midrange input impedance, @ 1 kΩ to 10 kΩ. If you connect a BJT CE amplifier to another BJT CE amplifier, the second amplifier will draw current from the first amplifier, reducing the gain of the first amplifier. In other words, the second amplifier "loads down" the first amplifier, changing its gain.

With JFETs we don't have that problem because the gate input impedance of a JFET is extremely high and $I_G = 0$ A. If you connect one JFET amplifier to a second JFET amplifier, the second amplifier draws no current from the first, so the gain of the first stage is unaffected.

The gain of a JFET also differs from that of a BJT. Remember, the gain of an amplifier is simply the output signal divided by the input signal; that is, how many times larger the output signal is than the input signal. Two out of three JFET configurations have a voltage gain, and thus the JFET can be used as a voltage amplifier. However, a JFET's voltage gain A_V is much lower than that of a comparable BJT's A_V.

A JFET amplifies an AC signal, but it needs a DC voltage (bias) to supply its operating power. Like the BJT, the most common and preferred method used to bias JFET amplifiers is called voltage-divider bias. In voltage-divider bias, the main DC power supply voltage is divided among various resistors to turn on a transistor so that it's ready for amplification. Figure 7-7

shows voltage-divider bias and the distribution of DC voltages in a typical JFET amplifier.

The main DC supply voltage for a JFET amplifier is called the V_{DD}, which is a positive voltage applied across the entire JFET with respect to ground. For Figure 7-7, the V_{DD} is 18 V. Recall that the main DC supply voltage for a BJT amplifier is called the V_{CC}.

As far as voltage distribution goes, it's as though you can divide a JFET in half—a left side and a right side. The blue dotted line in Figure 7-7 shows this division.

On the left side, the V_{DD} is divided between resistors R_1 and R_G. Resistor R_1's main job is to drop a good portion of the V_{DD}, and you can see that 9.9 V appears across R_1. The remainder of the V_{DD} (8.1 V) is dropped across resistor R_G. Resistor R_G is often called the biasing resistor because it applies the reverse voltage across the gate-to-source junction (V_{GS}) to control the drain current I_D.

On the right side of the JFET in Figure 7-7, the V_{DD} is divided among two resistors and the JFET itself. Resistor R_D is the drain resistor, and it has 1.4 V dropped across it. When the JFET is on, it is also behaving like a resistor, so it drops a voltage, which is called V_{DS}. In Figure 7-7, the JFET has 7 V appearing across it. Finally, resistor R_S, which is connected to the source, drops 9.6 V. This voltage across the source resistor is often called V_S.

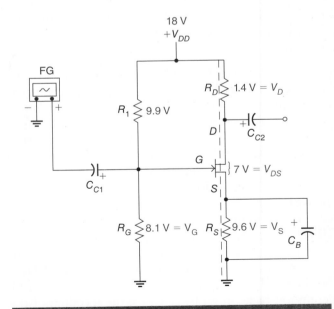

FIGURE 7-7 Voltage-divider bias and distribution of DC voltages in a JFET amplifier (© Cengage Learning 2012)

As with the BJT, all the DC voltages on the left side of the JFET add up to the total voltage, or V_{DD}. ($V_{R1} + V_G = V_{DD}$). Also, all the DC voltages on the right side of the transistor add up to the total voltage, or V_{DD}. ($V_D + V_{DS} + V_S = V_{DD}$). Thus, voltage-divider bias for JFETs is similar to voltage-divider bias for BJTs.

The JFET amplifier DC voltages are summarized below:

V_{DD} Positive drain supply voltage; supplies voltage for entire JFET.

V_G Voltage between the gate terminal and ground.

V_{GS} Voltage between the gate and source terminals; reverse biases the JFET.

V_{DS} Voltage between the drain and source (voltage drop across entire JFET).

V_S Voltage between the source and ground (voltage drop across source resistor).

7-6 COMMON-SOURCE AMPLIFIER

The three JFET amplifier configurations are the common-source, the common-drain, and the common-gate. Like the BJT, a JFET amplifier configuration is determined by the relationship between the AC input signal and the AC output signal. Remember, a bias, or DC voltage, supplies the operating power for a JFET, but a JFET amplifies an AC signal.

For example, if the input signal is applied to the gate terminal and the output signal appears at the drain terminal, then the configuration is called a **common-source amplifier**, and it will have some features similar to the BJT common-emitter. Using the *same* JFET, if the input signal is applied to the gate and the output signal appears at the source, then the configuration is called a **common-drain amplifier**, and it will have some features similar to the BJT common-collector. Again, using the *same* JFET, if the input signal is applied to the source and the output signal appears at the drain, then the configuration is called a **common-gate amplifier**, and it will have some features similar to the BJT common-base. In all three cases, the terminal of the JFET that doesn't touch the input signal or output signal determines the configuration *name*.

The common-source (CS) is the most widely used JFET amplifier configuration. Figure 7-8

shows an N-channel common-source amplifier with resistor values and capacitor values.

Notice that the AC input signal is applied to the gate terminal and the AC output signal appears at the drain terminal. The AC output signal at the drain is flipped over when compared to the AC input signal; this is called a 180° phase shift. In other words, the output signal is reaching its positive peak when the input signal is reaching its negative peak and vice versa. This 180° phase shift also occurs in the BJT common-emitter amplifier. Unlike the common-emitter, however, the common-source amplifier has an extremely high input impedance and only a low voltage gain.

Remember, the extremely high input impedance of the common source amplifier is desirable in amplifiers because it does not affect or change the gain of the JFET amplifier connected to its input; that is, the common-source amplifier will not load down any amplifier preceding it. This makes the common-source amplifier highly desirable in electronic systems that have multiple amplifier stages such as radio receivers.

In Figure 7-8, the V_{DD} supplies 18 V DC to the entire JFET with respect to ground. C_{C1} and C_{C2} are coupling capacitors that are used to bring the AC signal in and out of the JFET amplifier. The AC signal steals power from the V_{DD} to become larger, but then the AC signal doesn't want to have anything to do with the DC once it has gotten its gain. So, C_{C2} keeps any DC within the JFET amplifier stage.

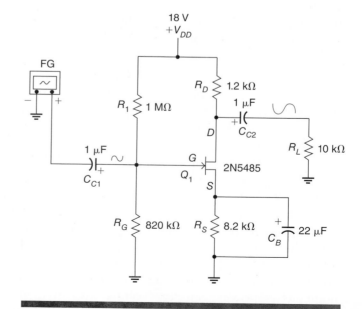

FIGURE 7-8 N-channel common-source amplifier (© Cengage Learning 2012)

Note that the source resistor R_S has a capacitor C_B in parallel with it. C_B stands for bypass capacitor. For maximum gain in a common-source amplifier, the AC signal should appear only at the drain terminal. If any of the AC signal leaks through the source terminal, it will lower the voltage gain of the amplifier. Thus, if any AC signal appears on the source terminal, the job of capacitor C_B is to bypass the source terminal and dump this AC to ground, where it will re-enter the gate terminal through resistor R_G. This bypass capacitor is needed to maintain the voltage gain of a common-source amplifier.

As we can see, the common-source JFET amplifier configuration in Figure 7-8 is similar in many ways to the common-emitter BJT amplifier that we analyzed in Chapter 5. They both use voltage-divider bias to supply the DC voltages for operation, and they both amplify AC signals. In Chapter 5, we were able to calculate the static voltage gain of a common-emitter amplifier on paper, and then we built the amplifier circuit and verified our calculations using measured voltages. Calculating the static voltage gain of a common-source JFET amplifier, however, is more complex and requires more than a calculator and paper. There are two main reasons for this complexity.

First, the common-emitter amplifier discussed in Chapter 5 uses a BJT, which is a current-controlled device. The biasing resistor V_B, or R_2, provides the I_B that controls the I_C, I_E, and ultimately the amplifier voltage gain. Also, the base-to-emitter voltage, or V_{BE}, that turns on the transistor is relatively constant: @ 0.3 V for a germanium transistor and @ 0.7 V for a silicon transistor.

The common-source amplifier, however, uses a JFET, which is a voltage-controlled device. The gate resistor R_G provides the negative V_{GS} that controls the I_D and ultimately the amplifier voltage gain. This gate-to-source voltage, or V_{GS}, however, is *not* constant: V_{GS} varies from JFET to JFET and from configuration to configuration.

Because of this varying V_{GS}, we are limited in our calculations for determining the static voltage gain (voltage gain without an AC signal applied) of the common-source amplifier. In the following examples, we'll be given values of V_{GS} to help us calculate the static voltage gain. For the lab activity, we'll have to build a common-source amplifier on the breadboard, measure its V_{GS}, and then use the measured value for our calculations of the static voltage gain. Then we'll apply an AC signal and measure and determine the dynamic voltage gain of the common-source amplifier.

Recall that every BJT has a DC current gain, or h_{FE}, or beta (β), which is an amplification rating listed on a spec sheet or in the *NTE* manual. The higher the h_{FE} for a BJT, the greater its amplification capability. This was represented in Chapter 4 by Equation 4-2:

$$\beta = \frac{I_C}{I_B}$$

Well, JFETs use a different amplification rating called the transconductance, or g_m. The **transconductance** of a JFET amplifier is the ratio of the change in I_D for a change in V_{GS}; that is, how much a change in V_{GS} will cause the I_D to change. The symbol for change is Δ, the Greek capital letter for *D*. The unit of transconductance is mho, which is ohm spelled backward, or siemens, S. The *NTE* manual lists μmhos for transconductance, and a spec sheet usually lists μmhos, although siemens (S) is gaining popularity as the unit of transconductance.

Spec sheets and the *NTE* manual use the abbreviation g_{fs} for transconductance, which is its value when $V_{GS} = 0$ V.

EQUATION 7-1

JFET transconductance

$$g_m = \frac{\Delta I_D}{\Delta V_{GS}}$$

g_m = transconductance in μmhos or μS

Δ = change in I_D or V_{GS}

While the BJT's h_{FE} can be measured directly using the h_{FE} setting/socket method, the transconductance of a JFET can't be measured directly, so you have to look it up in the *NTE* manual or on a spec sheet. Also, a JFET's transconductance varies with changing values of V_{GS}, so any calculations involving transconductance aren't exact; that is, they are ballpark values.

Equation 7-1 is the standard equation for determining transconductance, but we don't usually measure current values when designing or troubleshooting circuits; we measure voltages at certain locations in a circuit or stage (often called test points). Since the *NTE* manual lists the transconductance for a JFET when $V_{GS} = 0$ V (g_{fs}), we have to come up with a working value of transconductance when V_{GS} is at some other value than 0 V.

The following equation will help us calculate the working value of transconductance; that is, the transconductance when the operating DC voltages are applied to a JFET and V_{GS} is known. This equation is a variation of an equation used by William Shockley, a co-inventor of the BJT who also worked with JFETs.

EQUATION 7-2

JFET amplifier working value of transconductance

$$g_{mDC} = g_{fs}\left(1 - \frac{V_{GS}}{V_{GS\,(off)}}\right)$$

g_{mDC} = working value of transconductance in μmhos

g_{fs} = value of transconductance found in *NTE* manual (with $V_{GS} = 0\ V$)

V_{GS} = measured value of gate-to-source voltage

$V_{GS\,(off)}$ = gate-to-source cutoff voltage found in *NTE* manual

Now, we can begin the static voltage gain analysis of our CS amplifier in Figure 7-8. According to the *NTE* manual, the 2N5485 JFET (NTE 451) of Figure 7-8 has a transconductance or g_{fs} of 4000 μmhos, and its value of $V_{GS\,(off)}$ is −4 V. For Figure 7-8, the V_{GS} is −1.5 V. (I measured this value of V_{GS} when I built the circuit live.) First, we plug in our values.

1) $g_{mDC} = g_{fs}\left(1 - \dfrac{V_{GS}}{V_{GS(off)}}\right)$

$\quad = 4000\ \mu mhos\left(1 - \dfrac{-1.5\ V}{-4\ V}\right)$

$\quad = 4000\ \mu mhos\,(1 - 0.375) = 2500\ \mu mhos$

Now, we need one more equation to determine our CS amplifier's static gain.

EQUATION 7-3

JFET amplifier static voltage gain

$$A_V = (g_{mDC})(R_{OUT})$$

R_{OUT} = total resistance of drain resistor in parallel with load resistor

2) First, we find the value of R_{OUT}:

$$R_{OUT} = \frac{(R_D)(R_L)}{R_D + R_L} = \frac{(1.2\ k\Omega)(10\ k\Omega)}{1.2\ k\Omega + 10\ k\Omega}$$

$$= 1071.42\ \Omega$$

Then, we plug this into our final equation:

3) $A_V = (g_{mDC})(R_{OUT})$

$\quad = (2500\ \mu mhos)(1071.42\ \Omega) = 2.67$

Thus, the static voltage gain of the JFET amplifier of Figure 7-8 is 2.67, which means the AC output voltage signal should be *about* 2.67 times greater than the AC input voltage signal. Remember, these are ballpark calculations. JFETs have low voltage gains (<100), so this is a typical voltage gain for a JFET. Let's do some more examples before we build it live.

The static voltage gain of the JFET amplifier of Figure 7-9 is 4.79, which means the AC output voltage signal should be about 4.79 times greater than the AC input voltage signal. Again, JFETs have low voltage gains (<100), so this is a typical voltage gain for a JFET.

EXAMPLE 1

Situation

Calculate the static voltage gain for the JFET amplifier in Figure 7-9 using the *NTE* manual and Equations 7-2 and 7-3. The measured value of V_{GS} is −0.74 V.

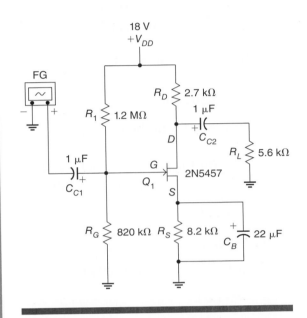

FIGURE 7-9   N-channel common-source amplifier (© Cengage Learning 2012)

Solution

According to the *NTE* manual, the 2N5457 JFET (NTE 457) of Figure 7-9 has a transconductance or g_{fs} of 3000 μmhos, and its value of $V_{GS\,(off)}$ is −6 V.

1) $$g_{mDC} = g_{fs}\left(1 - \frac{V_{GS}}{V_{GS\,(off)}}\right)$$

$$= 3000\ \text{μmhos}\left(1 - \frac{-0.74\ \text{V}}{-6\ \text{V}}\right)$$

$$= 3000\ \text{μmhos}\,(1 - 0.123)$$

$$= 2630\ \text{μmhos}$$

2) $$R_{OUT} = \frac{(R_D)(R_L)}{R_D + R_L} = \frac{(2.7\ \text{k}\Omega)(5.6\ \text{k}\Omega)}{2.7\ \text{k}\Omega + 5.6\ \text{k}\Omega}$$

$$= 1821.68\ \Omega$$

3) $$A_V = (g_{mDC})(R_{OUT})$$

$$= (2630\ \text{μmhos})(1821.68\ \Omega)$$

$$= 4.79$$

EXAMPLE 2

Situation

Calculate the static voltage gain for the JFET amplifier in Figure 7-10 using the *NTE* manual and Equations 7-2 and 7-3. The measured value of V_{GS} is −1.05 V.

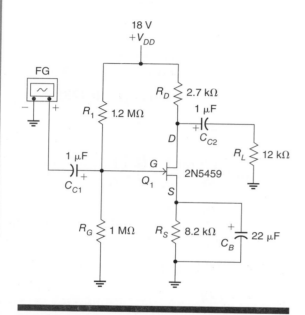

FIGURE 7-10   N-channel common-source amplifier (© Cengage Learning 2012)

Solution

According to the *NTE* manual, the 2N5459 JFET (NTE 459) of Figure 7-10 has a *maximum* transconductance of 6500 μmhos (when V_{GS} is 0 V), and its value of $V_{GS\,(off)}$ is −6 V.

1) $g_{mDC} = g_{fs}\left(1 - \dfrac{V_{GS}}{V_{GS\,(off)}}\right)$

$\quad = 6500\ \text{μmhos}\left(1 - \dfrac{-1.05\ \text{V}}{-6\ \text{V}}\right)$

$\quad = 6500\ \text{μmhos}\,(1 - 0.175)$

$\quad = 5362.5\ \text{μmhos}$

2) $R_{OUT} = \dfrac{(R_D)(R_L)}{R_D + R_L} = \dfrac{(2.7\ \text{k}\Omega)(12\ \text{k}\Omega)}{2.7\ \text{k}\Omega + 12\ \text{k}\Omega}$

$\quad = 2204.08\ \Omega$

3) $A_V = (g_{mDC})(R_{OUT})$

$\quad = (5362.5\ \text{μmhos})(2204.08\ \Omega)$

$\quad = 11.81$

The *maximum* static voltage gain of the JFET amplifier of Figure 7-10 is 11.81, which means the AC output voltage signal should be a maximum of 11.81 times greater than the AC input voltage signal. Remember, the transconductance, or g_{fs} value, listed in the *NTE* manual for the 2N5459 JFET is the maximum amount, so the measured static voltage gain will be less. Again, JFETs have low voltage gains (<100), so this is a typical voltage gain for a JFET.

Keep in mind that a JFET amplifier's static voltage gain is a ballpark calculation. The dynamic voltage gain of a JFET amplifier (as well as a BJT amplifier) is the actual measured gain. When a transistor is powered by DC and is amplifying an AC signal, then the dynamic

voltage gain can be determined. For the dynamic voltage gain, you measure the V_{IN}, the V_{OUT}, and then use Equation 5-3 from Chapter 5 to determine the A_V:

$$A_V = \frac{V_{OUT}}{V_{IN}}$$

In the following lab activity, we'll build a common-source amplifier on the breadboard, measure its V_{GS}, and then use the measured value for our calculations of the static voltage gain. Then we'll apply an AC signal and measure and determine the dynamic voltage gain of the common-source amplifier.

LAB ACTIVITY 7-3

Common-Source Amplifier

Materials, Equipment, and Parts:

- *NTE* catalog or Internet access, www.nteinc.com (*Look it up to hook it up.*)

- PC w/Multisim®, Electronics Workbench®, or SPICE.

- DMM with test leads.

- Dual trace oscilloscope w/BNC-to-alligator leads.

- 15 V DC voltage source.

- Function generator, 1 Hz to 10 kHz.

- Breadboard and connecting wires.

- 2N5485 or equivalent N-channel JFET.

- 1 µF electrolytic capacitor (2) and 22 µF electrolytic capacitor.

- 1.2 kΩ fixed resistor, 2.7 kΩ fixed resistor, 4.3 kΩ fixed resistor, 1 MΩ fixed resistor, and 1.2 MΩ fixed resistor.

Discussion Summary:

A JFET can be used as a common-source (CS) amplifier. The CS has an AC input signal applied to the gate terminal, and the AC output signal appears across the drain terminal. One feature of the CS is a 180° phase shift of the AC signal; the AC signal at the output terminal is flipped over when compared to the AC input signal. In other words, the output signal is reaching its positive peak when the input signal is reaching its negative peak and vice versa. Another feature of the CS is a low voltage gain, or A_V, of less than 100. In this lab we'll calculate the static voltage gain and then measure the dynamic voltage gain of the CS amplifier.

Procedure:

SAFETY FIRST. Eye protection should always be worn when working with live voltages. Before powering on a live circuit, always check with your instructor.

1 Write the part number of the JFET at your workstation. Then use the *NTE* catalog or visit the website www.nteinc.com to find and record the NTE replacement number and the diagram number.

Part number _____

NTE number _____

Diagram # _____

Transconductance (g_{fs}) _____

$V_{GS (off)}$ _____

[make it a negative (−) value if it's not listed as negative]

2 Draw the JFET transistor outline and label the drain, source, and gate.

(continues)

LAB ACTIVITY **7-3**

(continued)

3 Build the circuit of Figure 7-11 on your breadboard.

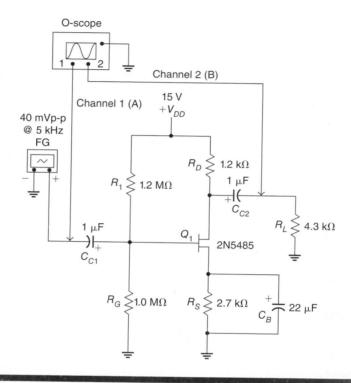

FIGURE 7-11   Common-source amplifier with 2N5485 JFET (© Cengage Learning 2012)

4 Connect the positive lead of the 15 V DC voltage source to the top end of resistors R_1 and R_D.

5 Connect the positive lead of the function generator to the negative side of capacitor C_{C1}.

6 Connect the positive lead of the oscilloscope's Channel 1 (A) to the negative side of capacitor C_{C1}. Connect the positive lead of the o-scope's Channel 2 (B) to the negative side of capacitor C_{C2}.

7 Connect the black lead of the function generator, the black lead of the 15 V DC voltage source, the black leads of the o-scope, a wire from the bottom of resistor R_G, and a wire from the bottom of resistor R_S to one point (hole) on the breadboard, which is ground.

8 Set the function generator to the sine wave setting and to an amplitude of 40 mV peak-to-peak. You may have to use your o-scope's Channel 1 volts/division setting to adjust the voltage to 40 mV.

LAB ACTIVITY **7 - 3**

9 Set the function generator for a frequency of 5 kHz.

10 Have the instructor check your circuit.

11 Power on only the 15 V DC voltage source. Connect the positive lead of your DMM to the gate of the JFET and the negative lead of your DMM to the source terminal of the JFET. Record the V_{GS}. It will be a negative value.

V_{GS} _____

Power off the DC voltage source and remove your DMM leads.

12 Using the values from Step 1, Step 11, and Figure 7-11, calculate and record the *static* voltage gain of the CS amplifier.

$$g_{mDC} = g_{fs}\left(1 - \frac{V_{GS}}{V_{GS(off)}}\right) =$$

$$R_{OUT} = \frac{(R_D)(R_L)}{R_D + R_L} =$$

$$A_V = (g_{mDC})(R_{OUT}) =$$

13 Power on the 15 V DC voltage source, the function generator, and the o-scope. Use your o-scope to ensure the AC input signal on Channel 1 is 40 mV peak-to-peak. You may have to adjust the Channel 1 volts/division setting and/or other o-scope settings. Record the value below.

V_{IN} _____

14 Measure and record the AC output voltage that appears on the o-scope's Channel 2. You may have to adjust the Channel 2 volts/division setting and/or other o-scope settings. Record the value below.

V_{OUT} _____

15 Use the equation below to calculate and record the dynamic voltage gain of the CS amplifier.

$$A_V = \frac{V_{OUT}}{V_{IN}} =$$

A_V _____

How does this measured A_V compare to the calculated static A_V of Step 12? Explain any differences.

(continues)

LAB ACTIVITY 7-3

(continued)

16 Set your o-scope's vertical mode switch to the dual setting. Adjust your o-scope settings so that the 40 mV input signal is on top and the V_{OUT} signal is directly below it. Are the waveforms mirror images; that is, does one peak positive at the same time the other peaks negative?

17 Power off the 15 V DC source, the function generator, and the o-scope. Remove capacitor C_B from the circuit. Power on the 15 V DC source, the function generator, and the o-scope. Use the equation below to recalculate and record the dynamic voltage gain of the CS amplifier.

$$A_V = \frac{V_{OUT}}{V_{IN}} =$$

A_V _____

How does this measured A_V compare to the measured A_V of Step 15? Explain any differences.

18 Have the instructor verify your results.

19 Build the circuit on Multisim®, Electronics Workbench®, or SPICE and then compare the results.

TABLE 7-1 BJT and JFET comparison and contrast (© Cengage Learning 2012)

BJT	JFET
• Current-controlled—I_B controls I_C	• Voltage-controlled—V_{GS} controls I_D
• Base to emitter is forward biased	• Gate to source is reverse biased
• V_{CC} is DC supply voltage	• V_{DD} is DC supply voltage
• $I_E = I_C + I_B$	• $I_D = I_S$, $I_G = 0$ A
• Cutoff: $I_C = 0$ A, V_{CE} = maximum (V_{CC} value)	• Cutoff or $V_{GS(off)}$: $I_D = 0$ A, $V_{GS} = V_P$ (V_{GS} and V_P have same value, but opposite polarity)
• Midpoint: $I_C \approx$ ½ maximum, $V_{CE} \approx$ ½ maximum (½ of V_{CC} value)	• Constant-current region: I_D determined by negative values of V_{GS}
• Saturation: totally on, I_C = maximum, $V_{CE} = 0$ V	• V_P, or pinch-off voltage (saturation): totally on, I_D = maximum, $V_{GS} = 0$ V
• Z_{IN} from Ω to kΩ	• Z_{IN} from kΩ to GΩ (varies w/config.)
• Appearance: 3 leads, various case styles	• Appearance: 3 leads, various case styles
• Part number begins with 2N	• Part number begins with 2N
• Pin outline—Look it up to hook it up	• Pin outline—Look it up to hook it up
• NPN most common	• N-channel most common
• Common-emitter	• Common-source
• Common-collector	• Common-drain
• Common-base	• Common-gate

Now that we've identified, tested, and built circuits with JFETs and BJTs, it's time to summarize their similarities and differences. Table 7-1 compares and contrasts BJTs and JFETs.

Notice the last three bullets, which list amplifier configurations. The JFET common-source amplifier is similar to the BJT common-emitter for two reasons: the AC input signal enters the gate (base for BJT) and exits from the drain (collector in BJT); the common-source amplifier shifts the input signal 180°; that is, the AC output signal at the drain is flipped over when compared to the AC input signal. However, the similarities end there. The common-emitter amplifier has midrange values for Z_{IN}, Z_{OUT}, and A_V; the common-source has a very high Z_{IN} and Z_{OUT} and a low A_V.

Likewise, the JFET common-drain has a BJT counterpart in the common-collector. Both the JFET common-drain and the BJT common-collector have high Z_{IN}, low Z_{OUT}, an $A_V \leq 1$, and are called followers.

Finally, the JFET common-gate amplifier is similar to the BJT common-base amplifier. Both the common-gate and the common-base have a low Z_{IN} and a high Z_{OUT}. But unlike the common-base, the common-gate has a low A_V.

We've covered the JFET in detail. Now, we'll look at the second category of FETs—the IGFET (Insulated Gate Field Effect Transistor)—and its most common representative, the MOSFET.

7-7 INTRODUCTION TO MOSFETs

The **MOSFET (Metal Oxide Semiconductor Field Effect Transistor)** (pronounced "MOSS-FET") was invented in 1966 by Robert W. Bower, who worked for Hughes Research Laboratories in Malibu, California. However, the patent for the MOSFET was issued to the Bell Labs team of Kerwin,

Klein, and Sarace in 1969. Both Bower and the Bell Labs teams helped spur the MOSFET industry.

INTERNET ALERT

Check the website www.bell-labs.com/history to see the many electronic and computer devices that were invented by employees of Bell Labs.

Like the BJT and JFET, the MOSFET is a transistor that can be used as a high-speed switch or as an amplifier. MOSFETs, JFETs, and BJTs often look the same, and again, the only way to tell the difference is *Look it up to hook it up.* Figure 7-12 shows three typical FET case styles and a static-resistant bag. Can you tell which one is the MOSFET?

The JFET is on the left. The middle and right transistors are MOSFETs. Like the JFET, the MOSFET is a voltage-controlled device with drain, gate, and source terminals. The JFET's gate terminal is connected directly to either an N-channel or a P-channel. The MOSFET's gate terminal, however, is separated or insulated from the rest of the housing. That's why the MOSFET is also called an Insulated Gate Field Effect Transistor, or **IGFET**.

For the past 40 years, the most common insulating material was made of silicon dioxide, a glasslike substance. This insulating oxide put the "O" in MOSFET. In the last couple of years, however, many companies like Intel and AMD have started using hafnium dioxide as the insulating material in MOSFET gates for CPUs. Hafnium dioxide provides better insulation and reduces power losses. The new design for MOSFET gates is called 45 nanometer technology, or 45 nm technology, because that's the size of the insulating region—only a few *atoms* wide!

The insulating material used in MOSFETs is much thinner than a hair; this thinness is good and bad. This narrow region in a MOSFET allows for incredibly fast switching on-off times, making the MOSFET the workhorse in computer CPUs. However, the insulating material in MOSFETs is so thin that it can be easily damaged by **electrostatic discharge (ESD)**, which is commonly called static electricity. Remember, 15,000 volts or more of static electricity (at a low current) can be generated by just walking across a carpet or rubbing your back against a cloth chair. Static electricity is a silent killer, more intense during cool, dry weather. As little as 10 V can break down the insulator and ruin a MOSFET. When a MOSFET's gate is destroyed by ESD, there's no telltale sign like a hiss followed by a wisp of smoke. The MOSFET just won't work.

Because of the ESD problem with MOSFETs, manufacturers often construct MOSFETs with one or two Zener diodes internally connected between the gate and source terminals. (Sometimes a Zener is also internally connected between the drain and source terminals.) The Zeners will start conducting at their rated voltage and divert any potentially damaging voltages away from the MOSFET insulating material. In the *NTE* manual, such MOSFETS will have the phrase "Zener Protected" or "Gate Protected" listed under the column "Description and Application." Also, manufacturers ship MOSFETs in silvery-black, static-resistant bags like the one shown in Figure 7-12. These bags protect the MOSFETs during shipping and handling. Computer parts such as motherboards, CPUs, and RAM often contain MOSFETs, and they are also shipped in static-resistant bags.

Many electronic bench technicians often wear a grounded wrist strap or use antistatic mats at their workstations to prevent ESD from destroying a MOSFET. (MOSFETs already connected in a circuit aren't affected by ESD.) If you don't have a grounded wrist strap or antistatic mat, perform the following steps when using MOSFETs. Before you pick up the MOSFET, touch the metal case of

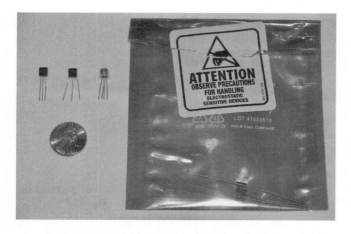

FIGURE 7-12 Typical FET case styles and a static-resistant bag (Courtesy of Tracy Grace Leleux)

the unit you are working on; this will discharge any static electricity from your body. Then, carefully pick up the MOSFET by the case—not the leads—and insert the MOSFET in the circuit. Finally, touch the metal case again just before you apply the soldering iron and solder to each lead. Ideally, you should use a grounded soldering iron. ESD rarely destroys BJTs or JFETs, but it's a good idea to get in the practice of grounding yourself before touching any out-of-circuit transistor.

Troubleshooting Tip

Touch the metal case of the unit you're working with before handling a MOSFET.

The two types of MOSFETs are D-MOSFETs and E-MOSFETs. The **D-MOSFET** is named for depletion mode and the **E-MOSFET** for enhancement mode. These modes, or methods, of operation will be discussed soon. We'll tackle the D-MOSFET first because it has a mode of operation that closely resembles the behavior of a JFET.

7-8 D-MOSFETs

Figure 7-13 shows simplified construction diagrams and schematic symbols for N-channel and P-channel D-MOSFETs. In Figure 7-13(a) we see the gate on the far left of the construction diagram. Gates were made of metal—usually aluminum—from the 1960s to the mid-1970s. Then the aluminum gate was replaced with a nonmetallic material called polysilicon. Now, manufacturers are using metal again for the gates of MOSFETs.

To the right of the gate is the insulating material, either silicon dioxide or hafnium dioxide. To the right of the insulating material is the N-channel, and to the right of the N-channel is the P-type material, called the substrate or base. The N-channel and P-type material make up the semiconductor part of the MOSFET. Thus, in Figure 7-13(a), from left to right we have Metal (gate)—Oxide (insulator)—Semiconductor (N-channel and P-type base)—Field—Effect—Transistor, or MOSFET.

Notice in the schematic symbol of Figure 7-13(a) that the gate terminal is not connected to the rest of

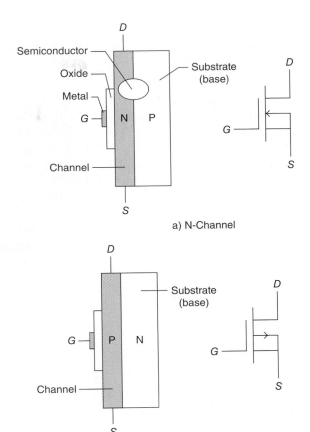

FIGURE 7-13 D-MOSFET construction diagrams and schematic symbols: a) N-channel and b) P-channel (© Cengage Learning 2012)

the MOSFET; this physical separation represents the insulating material. Since the gate is made of metal (a conductor) and since the N-type channel is also a conductor, we have two metals separated by an insulator. Sounds familiar? Yes, we have a capacitive effect between the gate and the N-channel.

Link to Prior Learning

A capacitor is two conductors, or metal plates, separated by an insulator, often called the dielectric.

Looking again at the schematic symbol in Figure 7-13(a), we see that the arrow connected to the source terminal is pointing iN for an N-channel D-MOSFET. For the P-channel D-MOSFET of Figure 7-13(b), the arrow is pointing out, and the channel is made of P-type material.

The D-MOSFET can be operated using one of two possible methods or modes: depletion mode or enhancement mode. The "D" in D-MOSFET stands for depletion. The D-MOSFET is apparently losing popularity—the 14th edition of the *NTE* manual only lists four D-MOSFETs out of 4700 electronic replacement parts, and those four have dual gates, which will be discussed later. Still, the D-MOSFET is versatile and worth investigating. Figure 7-14 shows the basic operation of the D-MOSFET in both depletion mode and

enhancement mode. Again, these circuits are for learning purposes only.

The D-MOSFET's **depletion mode** is similar to the operation of the JFET. Like the JFET, a D-MOSFET configured for depletion mode is totally on (maximum I_D) when the V_{GS} is 0 V, as shown in the top circuit of Figure 7-14(a).

If you look carefully at the top circuit of Figure 7-14(a), you will see that the gate-to-source junction is reverse biased by the 0 V DC voltage source on the left. This DC voltage on the left controls the gate-to-source voltage, or V_{GS}.

On the right side of the D-MOSFET, the positive end of the 90 V DC voltage source is connected

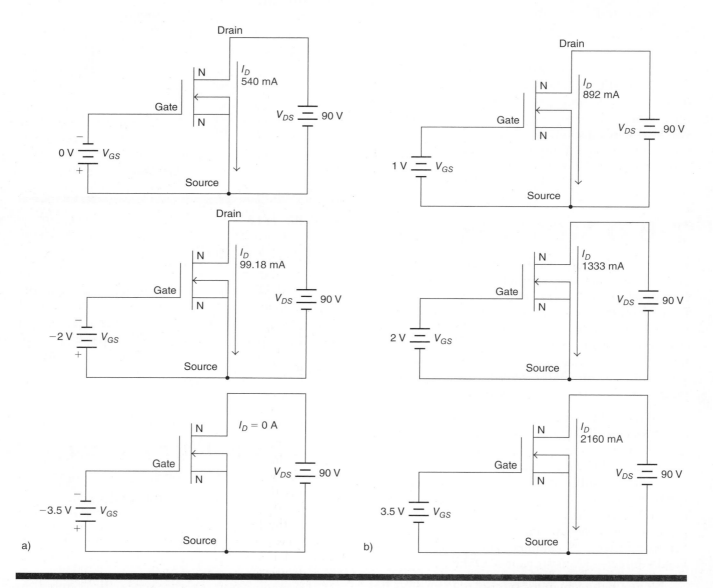

FIGURE 7-14 N-channel D-MOSFET modes of operation: a) Depletion mode and b) Enhancement mode
(© Cengage Learning 2012)

to the drain terminal, which in turn is connected to the N-channel, so the drain is reverse biased. This 90 V DC voltage source on the right controls the drain-to-source voltage, or V_{DS}. (This value may seem high compared to voltages we've used throughout this textbook so far, but the voltages and currents for Figure 7-14 are derived from a commercial D-MOSFET, part number DN1509.)

INTERNET ALERT

Check the website http://www.falstad.com/ circuit/e-index.html to see the operation of a MOSFET and other semiconductor circuits.

Link to Prior Learning

Although many transistor part numbers have the prefix 2N, different manufacturers use different prefixes. Again, *Look it up to hook it up.*

How exactly do the circuits of Figure 7-14(a) operate? The 90 V DC voltage source (V_{DS}) on the right causes conventional, or hole, current to flow down through the N-channel. If the DC voltage source on the left that controls the V_{GS} is set to 0 V as shown in the top circuit of Figure 7-14(a), then the drain current (I_D) is maximum, which in this case is 540 mA. Like the JFET, a D-MOSFET operating in depletion mode is totally on (maximum I_D) when the V_{GS} is 0 V. Current is at maximum because the depletion region is small where the gate's insulating material meets the N-channel. No current flows into the gate terminal because it is reverse biased. Also, the insulating material would prevent any electrons from moving there.

In the middle circuit of Figure 7-14(a), the V_{GS} has been changed to −2 V. By making V_{GS} more negative, we've increased the reverse bias. As reverse bias increases, the depletion region reduces or depletes the N-channel; that's why this mode of operation is called depletion mode. This increased depletion region reduces the drain current to 99.18 mA.

In the bottom circuit of Figure 7-14(a), the V_{GS} has been changed to −3.5 V. The reverse bias and depletion region are at maximum, so no drain current flows. This condition is like cutoff in a BJT, and it is called the gate-to-source cutoff voltage, or $V_{GS\ (off)}$. In summary, when the D-MOSFET is configured for depletion mode, its operation is very similar to that of the JFET.

The D-MOSFET can also be operated in **enhancement mode**. In enhancement mode, the MOSFET's gate-to-source junction is *forward biased.* Figure 7-14(b) shows the same D-MOSFET (DN1509) configured for enhancement mode.

If you look carefully at the top circuit of Figure 7-14(b), you will see that the gate-to-source junction is now forward biased by the 1 V DC voltage source on the left. *This is the first time that we've talked about a forward bias on the gate-to-source junction.* Again, this DC voltage on the left controls the gate-to-source voltage, or V_{GS}.

On the right side of the D-MOSFET, the positive end of the 90 V DC voltage source is connected to the drain terminal, which in turn is connected to the N-channel, so the drain is reverse biased. Again, this 90 V DC voltage source on the right controls the drain-to-source voltage, or V_{DS}. So, the 90 V DC voltage source hasn't changed for enhancement mode.

How do the circuits of Figure 7-14(b) operate? The 90 V DC voltage source (V_{DS}) on the right causes conventional, or hole, current to flow down through the N-channel. When the DC voltage source on the left that controls the V_{GS} is set to positive 1 V as shown in the top circuit of Figure 7-14(b), the drain current (I_D) increases beyond the value that was the maximum for depletion mode. This is because the forward bias of the positive V_{GS} reduces the depletion region *and* expands the N-channel. The I_D of Figure 7-14(b) is now 892 mA as compared to 540 mA. Also, despite the fact that the gate-to-source junction is now forward biased, no current flows into the gate terminal because the insulating material prevents any electrons from going there.

In the middle circuit of Figure 7-14(b), the V_{GS} has been changed to positive 2 V. By making

V_{GS} more positive, we've increased the N-channel and I_D even more so that the value is now 1,333 mA.

Finally, in the bottom circuit of Figure 7-14(b), the V_{GS} has been changed to positive 3.5 V. The I_D is now 2,160 mA, four times greater than the maximum drain current for depletion mode!

The results of Figure 7-14(b) show that much higher values of I_D can be achieved when operating the same D-MOSFET in enhancement mode versus depletion mode. This is the main reason for the increasing popularity of enhancement-mode MOSFETs. One last point about D-MOSFETs is worth mentioning. Whether operated in depletion mode or enhancement mode, a D-MOSFET has no gate current because of the insulating material that separates the gate from the channel. This means that the Z_{IN} of a D-MOSFET is even higher than that of a JFET. A JFET has a Z_{IN} in the kΩ to GΩ range; a D-MOSFET could have a Z_{IN} in the TΩ range! This makes MOSFETs invaluable in low-current, low-power applications because their gate inputs don't draw any current. MOSFETs are used extensively in digital and computer circuitry including CMOS ICs and CPU chips manufactured by Intel and AMD.

7-9 E-MOSFETs

Although the D-MOSFET was the first MOSFET used in commercial applications, its peculiar versatility led to the development of the stand-alone E-MOSFET, which can only be operated in enhancement mode. The E-MOSFET has almost completely replaced the D-MOSFET in commercial applications.

Figure 7-15 shows simplified construction diagrams and schematic symbols for N-channel and P-channel E-MOSFETs. The construction diagram of Figure 7-15(a) closely resembles the D-MOSFET construction diagram of Figure 7-13(a). From left to right are gate, insulator, and channel. The major difference is the channel; it no longer connects the drain terminal to the source terminal. This broken channel is also shown in the schematic symbol of Figure 7-15(a) by a broken line. The schematic symbol in Figure 7-15(a) also has the arrow connected to the source terminal pointing iN for an N-channel E-MOSFET. For the P-channel E-MOSFET of Figure 7-15(b), the arrow is pointing out, and the channel is made of P-type material.

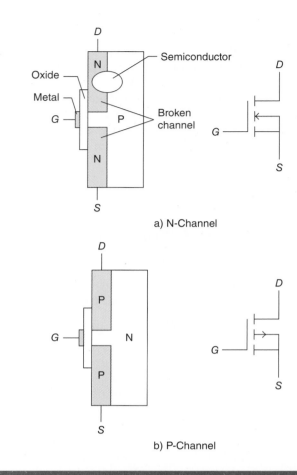

a) N-Channel

b) P-Channel

FIGURE 7-15 E-MOSFET construction diagrams and schematic symbols: a) N-channel and b) P-channel (© Cengage Learning 2012)

Because the channel is incomplete, the E-MOSFET can be operated *only* in enhancement mode; that is, the V_{GS} must be positive to turn on the MOSFET. This differs from the JFET and D-MOSFET depletion-mode operation in which they are totally on (maximum I_D) when the V_{GS} is 0 V.

Figure 7-16 shows the operation of an N-channel E-MOSFET with a constant 25 V value of V_{DS} and varying values of V_{GS}. The E-MOSFET's part number is BS170 (NTE 490).

When V_{GS} is 0 V as shown in Figure 7-16(a), the channel is incomplete, and no I_D flows. When V_{GS} is positive 2 V as shown in Figure 7-16(b), electrons are attracted from the P-type material to the gate, forming a bridge between the drain and source. The I_D is now 1.85 mA. This 2 V value for the BS170 E-MOSFET is significant because it's the voltage level at which the channel bridge is formed and the E-MOSFET turns on. This value is often

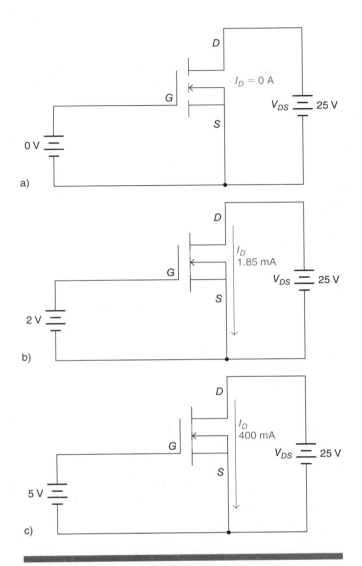

FIGURE 7-16 N-channel E-MOSFET operation with different values of V_{GS}: a) V_{GS} @ 0 V, b) V_{GS} @ 2 V, and c) V_{GS} @ 5 V (© Cengage Learning 2012)

called the **threshold voltage** and is listed on spec sheets as $V_{GS\,(Th)}$. Figure 7-17 shows a specification sheet for the BS170 E-MOSFET, NTE replacement number 490. This 2 V value is listed as the "Typ" or typical value for $V_{GS\,(Th)}$. The $V_{GS\,(Th)}$ for an E-MOSFET is like the V_{BE} for a BJT. Remember, a BJT made of silicon needs @ 0.7 V across its base-to-emitter junction to turn on.

When V_{GS} is increased to positive 5 V as shown in Figure 7-16(c), the channel bridge enlarges, and the I_D increases to 400 mA. For a 3 V change, this is an incredibly large increase in I_D. This is one of the main reasons why the E-MOSFET has become so popular in commercial applications. Another reason is the fast switching time. The turn-on/turn-off time for the BS170 is 10 nS.

Like the D-MOSFET, the input impedance for an E-MOSFET is also in the TΩ range, so its gate draws no current.

7-10 ADDITIONAL INSULATED GATE TRANSISTORS

At the beginning of this chapter, we discussed how the gate-insulator-channel makeup of a MOSFET is similar to a capacitor. This capacitive effect is not a consideration at low frequencies, but at higher frequencies, the small leakage current between the gate and channel can result in increased power consumption. The **MuGFET (Multiple Gate Field Effect Transistor)** provides a solution to this problem by using two or more gates to reduce the surface area and thus the overall capacitance. Although MuGFETs aren't nearly as common as other discrete MOSFETs, they are gaining more attention in CPU manufacturing. Figure 7-18 shows the schematic symbol for one example of a MuGFET, the dual-gate MOSFET.

The dual-gate MOSFET is a four-terminal MOSFET used in very high-frequency (VHF) radio receivers. By using two gate terminals, the overall capacitive effect is reduced because there is less surface area between the metallic gate and the conducting channel. Thus, the dual-gate MOSFET provides for both stable tuning of radio frequencies and excellent gain control.

The final insulated gate transistor that we'll look at is a hybrid transistor that's gaining popularity in semiconductor applications—the IGBT. Up until now we've been talking about the advantages of MOSFETs when compared to JFETs and BJTs—faster on/off times, less power consumption, and reduced loading effects. However, the MOSFETs we've looked at so far had drain currents in the mA to low amp range and power dissipation in the mW range. These are typically called small signal MOSFETs.

Remember, BJTs were invented in the 1940s, JFETs appeared in the 1950s, and MOSFETs became commercially available in the early 1970s. For many years, MOSFET manufacturers were playing catch-up in developing MOSFETs for use in high-current, high-power applications, those involving more than a 100 amps or watts. The resulting power MOSFETs can handle a little

ELECTRONICS, INC.
44 FARRAND STREET
BLOOMFIELD, NJ 07003
(973) 748–5089

NTE490
MOSFET
N–Ch, Enhancement Mode
High Speed Switch

Absolute Maximum Ratings:

Drain–Source Voltage, V_{DS} 60V

Gate–Source Voltage, V_{GS} . . . . ±20V

Drain Current (Note 1), I_D 500mA

Total Device Dissipation (T_A = +25°C), P_D 350mW

Operating Junction Temperature Range, T_J . . . . −55° to +150°C

Storage Temperature Range, T_{stg} . . . . −55° to +150°C

Note 1. The Power Dissipation of the package may result in a lower continuous drain current.

Electrical Characteristics: (T_A = +25°C unless otherwise specified)

Parameter	Symbol	Test Conditions	Min	Typ	Max	Unit
OFF Characteristics						
Drain–Source Breakdown Voltage	$V_{(BR)DSS}$	V_{GS} = 0, I_D = 100μA	60	90	–	V
Gate Reverse Current	I_{GSS}	V_{GS} = 15V, V_{DS} = 0	–	0.01	10	nA
ON Characteristics (Note 2)						
Gate Threshold Voltage	$V_{GS(Th)}$	V_{DS} = V_{GS}, I_D = 1mA	0.8	2.0	3.0	V
Static Drain–Source ON Resistance	$r_{DS(on)}$	V_{GS} = 10V, I_D = 200mA	–	1.8	5.0	Ω
Drain Cutoff Current	$I_{D(off)}$	V_{DS} = 25V, V_{GS} = 0	–	–	0.5	μA
Forward Transconductance	g_{fs}	V_{DS} = 10V, I_D = 250mA	–	200	–	mmhos
Small–Signal Characteristics						
Input Capacitance	C_{iss}	V_{DS} = 10V, V_{GS} = 0, f = 1MHz	–	–	60	pF
Switching Characteristics						
Turn–On Time	t_{on}	I_D = 200mA	–	4	10	ns
Turn–Off Time	t_{off}	I_D = 200mA	–	4	10	ns

Note 2. Pulse Test: Pulse Width ≤ 300μs, Duty Cycle ≤ 2%.

FIGURE 7-17   Specification sheet for BS170 E-MOSFET, NTE 490 (Courtesy of NTE Electronics, Inc.)

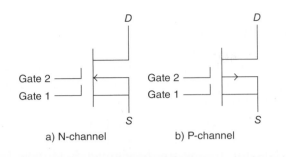

FIGURE 7-18 Schematic symbols for a dual-gate MOSFET: a) N-channel and b) P-channel (© Cengage Learning 2012)

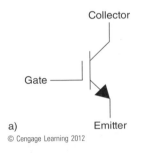

© Cengage Learning 2012

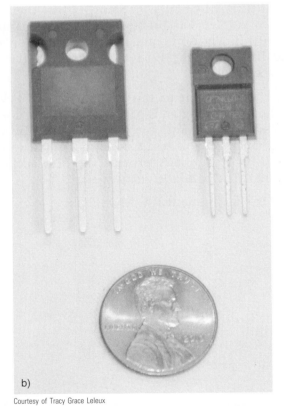

Courtesy of Tracy Grace Leleux

FIGURE 7-19 Insulated Gate Bipolar Transistor: a) Schematic symbol and b) Two IGBTs

more than 100 amps and up to 300 watts, and they are used in applications such as power supplies, low-voltage motor controllers, and antilock braking systems, to name a few. Now, MOSFET manufacturers have been focusing on making MOSFETs smaller and more power efficient, and MOSFETs have found their niche in the microcontroller and microprocessor industries.

In the 1980s, The **IGBT (Insulated Gate Bipolar Transistor)** (pronounced "IG-BET") came along to fulfill the role once dominated by power BJTs and JFETs—a fast-switching transistor capable of handling large current and power requirements. The IGBT is a three-terminal device that is a cross between a BJT and MOSFET in terms of construction and operation. Figure 7-19 shows the schematic symbol for the IGBT and a photo of two IGBTs.

Notice the terminals are marked collector, gate, and emitter. The insulated gate on the IGBT gives it the extremely high input impedance and high-speed turn-on capability of the MOSFET. An additional layer of P-type material gives the IGBT the ability to handle the high current and power requirements of the traditional power BJTs and power MOSFETs. The development of the IGBT began in the 1980s and continues through today, and it is becoming the transistor of choice for high-power systems that use hundreds of amps, volts, and watts. The IGBT is used in many applications including electric vehicles, hybrid cars, trains, stereos, high-voltage air conditioners, and uninterruptible power supplies for network servers.

7-11 IDENTIFYING AND TESTING E-MOSFETs

The easiest and most accurate way to test MOSFETs is to use a semiconductor tester such as a curvet tracer. However, you can still do a fairly accurate test using your DMM. Before you touch a MOSFET, ensure that you're wearing a grounded wrist strap or that you have an antistatic mat at your workstation to prevent ESD from destroying the MOSFET. If you don't have a grounded wrist strap or antistatic mat, touch the metal case of the unit you are working on, or touch some large metallic object. Always pick up the MOSFET by the case—not the leads.

To test an N-channel E-MOSFET, set your meter's rotary selector switch to the diode setting.

Then touch or clip the red (positive) lead of the meter to the drain terminal of the MOSFET, and touch or clip the black (negative) lead of the meter to the source terminal of the MOSFET. The drain and source channels are not physically connected, so the meter should display infinite, which may be represented by the infinity symbol (∞) or "OL" (overload) for a good N-channel E-MOSFET.

For the second check, touch or clip the red (positive) lead of the meter to the gate terminal of the E-MOSFET, and touch or clip the black (negative) lead of the meter to the source terminal of the E-MOSFET. This forward biases the gate-to-source junction and creates a bridge between the two N-channels. The meter will still show an infinite reading, indicating the gate is not shorted to the source.

For the third check, touch or clip the red (positive) lead of the meter to the drain terminal of the E-MOSFET, and touch or clip the black (negative) lead of the meter to the source terminal of the E-MOSFET. Since the source and drain are now connected to each other, the measurement should be a low voltage drop (@ 0 V) for a good N-channel E-MOSFET.

Next, touch or clip the red (positive) lead of the meter to the source terminal of the E-MOSFET, and touch or clip the black (negative) lead of the meter to the gate terminal of the E-MOSFET. This reverse biases the gate-to-source junction and collapses the bridge between the two N-channels. The meter will still show an infinite reading, indicating the gate is not shorted to the source.

Finally, verify that the bridge between the drain and source has collapsed. Touch or clip the red (positive) lead of the meter to the drain terminal of the MOSFET, and touch or clip the black (negative) lead of the meter to the source terminal of the MOSFET. The drain and source channels are not connected, so the meter should display infinite, which may be represented by the infinity symbol (∞) or "OL" (overload) for a good N-channel E-MOSFET.

In summary, E-MOSFET testing using the diode and meter leads requires five checks. The results for testing an N-channel E-MOSFET are summarized below:

1) Positive on drain and negative on source—display should be infinite or OL.

2) Positive on gate and negative on source—display should be infinite or OL.

3) Positive on drain and negative on source—display should be @ 0 V.

4) Positive on source and negative on gate—display should be infinite or OL.

5) Positive on drain and negative on source—display should be infinite or OL.

Notice that there is only one numerical reading and four infinite readings for a good N-channel E-MOSFET—in Step 3 after the gate-source voltage has bridged the two N-channels. If the ratio is different—two and three or zero and five—then the E-MOSFET is defective and must be replaced.

The limitation to this diode-setting/five-check method for testing an E-MOSFET is that it doesn't necessarily prove that the gate is intact. However, it is a fairly reliable test. Curve tracers and specialized MOSFET testers provide the best accuracy. Again, *when in doubt, swap it out.*

Let's test some real E-MOSFETs.

Identifying and Testing E-MOSFETs

Materials, Equipment, and Parts:

- *NTE* catalog or Internet access, www.nteinc.com

- Grounded wrist strap or antistatic mat.

- Digital Multimeter (DMM) with test leads.

- Two N-channel E-MOSFETs such as BS170, 2N7000, IRF510, IRF620, or equivalent.

Discussion Summary:

E-MOSFETs can go bad (become shorted or open). Technicians need to be able to identify the polarity of an E-MOSFET; the material it's made of (N-channel or P-channel); and the leads (source, gate, and drain) to test the transistor correctly. In this lab activity, you will use the diode-setting/five-check method for testing an E-MOSFET.

Procedure:

1 For each transistor at your workstation, use the *NTE* catalog to research and record the following information for each of the three transistors. For the first transistor, Steps 1 and 3 have been completed to serve as an example.

	1	2	3
Part #	BS170	_____	_____
NTE #	490	_____	_____
Polarity	E-MOSFET	_____	_____
Material	N-channel	_____	_____
Description & Application	enhancement mode, for switching applications	_____	_____
Case Style	TO92	_____	_____
Diagram #	9f	_____	_____

2 Go to the diagram page for each transistor and then draw the transistor outline (TO) for each transistor. Label the source, gate, and drain terminals.

1   2   3

(continues)

LAB ACTIVITY 7-4

(continued)

3 Use your DMM's diode setting and meter leads to check each transistor. Record the results. For the first volt reading, the positive lead (+) of the DMM is connected to the drain (D), and the negative lead (−) of the DMM is connected to the source terminal (S). Perform the remaining checks in this manner.

		1			2			3
+	−	Volt reading	+	−	Volt reading	+	−	Volt reading
D	S	OL	D	S	_____	D	S	_____
G	S	OL	G	S	_____	G	S	_____
D	S	0.011 V	D	S	_____	D	S	_____
S	G	OL	S	G	_____	S	G	_____
D	S	OL	D	S	_____	D	S	_____
Good?		Yes			_____			_____

4 Out of five checks, how many infinite readings should a good N-channel E-MOSFET measure?

5 Have the instructor check your results.

CHAPTER SUMMARY

The Field Effect Transistor is a three-terminal semiconductor device that is used as a high-speed switch or as an amplifier. FETs are voltage-controlled devices whose operation depends on the movement of either electrons *or* holes.

The two major categories of FETs are Junction Field Effect Transistors (JFETs) and Insulated Gate Field Effect Transistors (IGFETs). The most common IGFET is the Metal Oxide Semiconductor Field Effect Transistor (MOSFET). JFETs and MOSFETs differ in construction, but they are both voltage-controlled devices.

JFETs are constructed in one of two ways: N-channel or P-channel. The N-channel JFET is most common. The JFET has three terminals: drain, gate, and source. For the N-channel JFET, the drain terminal and the source terminal are connected to the same N-type material that has a large number of excess electrons. The gate terminal is connected to P-type material that has a small number of excess holes. The channel connecting the drain to the source is made of N-type material, so that's why it's called an N-channel JFET. A technician must be able to identify the type of JFET and each of the leads to test it correctly. JFET part numbers often have the prefix 2N, and each JFET has a manufacturer specification (spec) sheet.

For testing purposes, an N-channel JFET is really just one PN junction diode, and this diode can be measured from either the gate lead to the drain lead or from the gate lead to the source lead. A JFET can be tested using the diode-setting/six-check method. For a good JFET, there are four numerical readings and two infinite readings; these results are the opposite of a BJT.

A JFET operates with a reverse bias on the gate-to-source junction, called the V_{GS}, and a reverse bias on the drain-to-source junction, called the V_{DS}. When V_{GS} is 0 V, the JFET is totally on, and I_D is at maximum. If a point is reached where increasing the V_{DS} causes no more increase in I_D, the condition is similar to saturation in a BJT, and it is called (oddly) the pinch-off voltage, or V_P. As V_{GS} is made more negative, the reverse bias increases, and I_D is reduced or depleted. The point at which a negative V_{GS} stops the flow of I_D is called the gate-to-source cutoff voltage, or $V_{GS \text{ (off)}}$. A JFET has no I_G because the gate terminal is reverse biased. The gate-source junction is never forward biased in a circuit because a gate current greater than 50 mA can destroy a JFET.

Like a BJT, a JFET can be used as a high-speed switch or as an amplifier. Without an AC signal applied, a JFET is on, and I_D flows. The negative alternation of an AC voltage from a function generator reverse biases a JFET, stops the flow of I_D, and turns off any load connected to it, such as an LED.

A JFET is also used as an amplifier. The gate input impedance is incredibly high—in the MΩ to GΩ range. This is an advantage in multistage amplifiers because the JFET won't load the stage that comes before it. Two out of three JFET configurations have a voltage gain, so the JFET can be used as a voltage amplifier. However, a JFET's voltage gain A_V is lower than that of a comparable BJT's A_V.

Like the BJT, the most common and preferred method used to bias JFET amplifiers is called voltage-divider bias. The three JFET amplifier configurations are the common-source, the common-drain, and the common-gate. The common-source has some features similar to the BJT common-emitter, the common-drain has some features similar to the BJT common-collector, and the common-gate has some features similar to the BJT common-base.

The common-source (CS) is the most widely used JFET amplifier configuration. A JFET's transconductance, or g_m, is the ratio of the change in I_D for a change in V_{GS}; that is, how much a change in V_{GS} will cause the I_D to change. The unit of transconductance is the mho, which is ohm spelled backward, or siemens, S.

The Metal Oxide Semiconductor Field Effect Transistor (MOSFET) is a transistor that can be used as a high-speed switch or as an amplifier. Like the JFET, the MOSFET is a voltage-controlled device with drain, gate, and source terminals. However, the MOSFET's gate terminal is insulated from the rest of the housing. The narrow insulated region in a MOSFET allows for incredibly fast switching on/off times, making the MOSFET the workhorse in computer CPUs. However, the insulating material in MOSFETs is so thin that it can be easily damaged by ESD, so precautions must be taken when working with MOSFETs. Static-resistant bags, grounding straps, and anti-static mats help protect MOSFETs from ESD.

The two types of MOSFETs are D-MOSFETs and E-MOSFETs. D-MOSFET can be operated in depletion mode or enhancement mode. Depletion mode is similar to the operation of the JFET: I_D is at maximum when the V_{GS} is 0 V. By making V_{GS} more negative, reverse bias increases, and I_D decreases. In enhancement mode, the MOSFET's gate-to-source junction is *forward biased*, which increases I_D. The Z_{IN} of a D-MOSFET is in the TΩ range.

The E-MOSFET can only be operated in enhancement mode; that is, the V_{GS} must be positive to turn on the MOSFET. The voltage level at which an E-MOSFET forms a channel bridge and turns on is called the threshold voltage, $V_{GS\ (Th)}$.

Additional FETs include the Multiple Gate Field Effect Transistor (MuGFET), used for high-frequency applications, and the Insulated Gate Bipolar Transistor (IGBT), used for high-current, high-power applications such as stereo amplifiers, HVAC systems, and electric vehicles.

CHAPTER EQUATIONS

JFET transconductance

(Equation 7-1)

$$g_m = \frac{\Delta I_D}{\Delta V_{GS}}$$

g_m = transconductance in µmhos or µS

Δ = change in I_D or V_{GS}

JFET amplifier working value of transconductance

(Equation 7-2)

$$g_{mDC} = g_{fs}\left(1 - \frac{V_{GS}}{V_{GS(off)}}\right)$$

g_{mDC} = working value of transconductance in µmhos

g_{fs} = value of transconductance found in *NTE* manual (with $V_{GS} = 0$ V)

V_{GS} = measured value of gate-to-source voltage

$V_{GS(off)}$ = gate-to-source cutoff voltage found in *NTE* manual

JFET amplifier static voltage gain

(Equation 7-3)

$$A_V = (g_{mDC})(R_{OUT})$$

R_{OUT} = total resistance of drain resistor in parallel with load resistor

CHAPTER REVIEW QUESTIONS

Chapter 7-1

1. What do the letters in JFET stand for?

2. The BJT is a _____-controlled device, and the JFET is a _____-controlled device.

3. At first glance, how would you distinguish a BJT from a JFET?

Chapter 7-2

4. Draw and label the schematic symbol for an N-channel JFET.

5. What terminal of a JFET is similar to the base terminal of a BJT?

6. Is a transistor with the part number 2N4222 an N-channel or a P-channel JFET?

7. For proper operation, the gate-to-source junction of a JFET must be _____ biased.

8. What could happen if a forward bias greater than 50 mA is applied to the gate-to-source junction of a JFET?

9. When testing an N-channel JFET using the diode-setting/six-check method, your DMM measures infinite on three of the readings. Is the JFET good or bad?

Chapter 7-3

10. What is the difference between pinch-off voltage (V_P) and $V_{GS\,(off)}$ for a JFET?

11. At what value of V_{GS} is I_D maximum in a JFET?

12. At what value of V_{GS} is I_D minimum in a JFET?

13. What is the $V_{GS\,(off)}$ for a JFET whose part number is 2N5457?

14. If the $V_{GS\,(off)}$ value for a JFET is –4 V, what would the pinch-off voltage (V_P) value be?

15. Is the V_{GS} in a JFET usually a positive or a negative value?

Chapter 7-4

16. What are the two general applications for a JFET?

Chapter 7-5

17. The input impedance of a JFET is in the MΩ to GΩ range. Does this input impedance provide an advantage or disadvantage when using the JFET as an amplifier?

18. The main DC supply voltage for a BJT is called the V_{CC}. What is the main DC supply voltage for an FET called?

19. Voltage-divider bias can be used for a JFET amplifier circuit. What is the V_{DS} in a JFET amplifier?

Chapter 7-6

20. What does the transconductance rating of a JFET mean?

21. Calculate the g_{mDC}, R_{OUT}, and A_V for the circuit of Figure 7-20 when the following values are given: $g_{fs} = 3000$ µmhos; V_{GS} is -2.15 V; and $V_{GS\ (off)} = -6$ V.

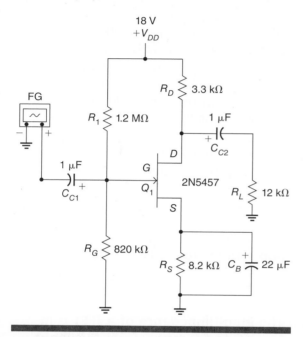

FIGURE 7-20   N-channel common-source amplifier (© Cengage Learning 2012)

$$g_{mDC} = g_{fs}\left(1 - \frac{V_{GS}}{V_{GS(off)}}\right) =$$

$$R_{OUT} = \frac{(R_D)(R_L)}{R_D + R_L} =$$

$$A_V = (g_{mDC})(R_{OUT}) =$$

22. Which JFET configuration has a voltage gain ≤ 1?

23. Which JFET configuration has the lowest input impedance?

24. Which JFET configuration inverts the AC input signal 180°?

25. Which JFET configuration is similar to the BJT common-emitter?

Chapter 7-7

26. What do the letters in MOSFET stand for?

27. Name one advantage of a MOSFET's insulating region between the gate and channel.

28. Name two methods to prevent ESD from damaging a MOSFET.

Chapter 7-8

29. Draw and label the schematic symbol for an N-channel D-MOSFET.

30. Which MOSFET can be operated in either depletion mode or enhancement mode?

31. In which MOSFET operating mode is the gate-to-source junction reverse biased, depletion mode or enhancement mode?

32. In which MOSFET operating mode is the gate-to-source junction forward biased, depletion mode or enhancement mode?

33. Which transistor type has the highest input impedance, a BJT, JFET, or MOSFET?

Chapter 7-9

34. Which MOSFET must be operated in enhancement mode?

35. What is the name for the voltage level at which an E-MOSFET forms a channel bridge and turns on?

36. **Which transistor type has the fastest turn-on time, a BJT, JFET, or E-MOSFET?**

Chapter 7-10

37. **Name three applications for an IGBT.**

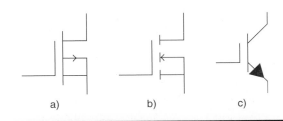

a)   b)   c)

FIGURE 7-21   Insulated gate transistors' schematic symbols (© Cengage Learning 2012)

38. **Which schematic symbol represents an N-channel E-MOSFET in Figure 7-21?**

39. **Which schematic symbol represents an IGBT in Figure 7-21?**

40. **Which schematic symbol represents a P-channel D-MOSFET in Figure 7-21?**

Advanced Semiconductor Devices and Applications

Operational Amplifiers

Upon completion of this chapter, you should be able to:

- Define operational amplifier (op amp) and explain its operating characteristics such as gain, input impedance, and output impedance.
- Draw and label the pin configuration for an LM741CN op amp.
- Draw and label the schematic symbol for a 5-pin op amp.
- Distinguish between the open-loop voltage gain and the closed-loop voltage gain of an op amp.
- Draw an op amp voltage follower circuit.
- Determine the static voltage gain and output voltage for an op amp non-inverting amplifier.

- Build an op amp non-inverting amplifier circuit to calculate and measure its dynamic voltage gain.
- Determine the static voltage gain and output voltage for an op amp inverting amplifier.
- Build an op amp inverting amplifier circuit to calculate and measure its dynamic voltage gain.
- Calculate the input voltage difference, static voltage gain, and output voltage for an op amp differential amplifier.
- Calculate the output voltage for an op amp summing amplifier.
- Build an op amp summing amplifier to calculate and measure its dynamic voltage gain.

MATERIALS, EQUIPMENT, AND PARTS

Materials, equipment, and parts needed for the lab experiments in this chapter are listed below:

- *NTE* catalog or Internet access, www.nteinc.com
- PC w/Multisim®, Electronics Workbench®, or SPICE.
- DMM with test leads.
- Dual trace oscilloscope w/BNC-to-alligator leads.
- Positive DC voltage source, 5 V.
- Positive DC voltage source, 0–20 V.
- Negative DC voltage source, 0–20 V.
- Function generator, 1 kHz to 10 kHz.
- Breadboard and connecting wires.
- LM741CN, UA741CN, or equivalent op amp IC.
- 1 kΩ fixed resistor (2), 2.7 kΩ fixed resistor, 3.9 kΩ fixed resistor, 4 kΩ fixed resistor, 5 kΩ fixed resistor, 6.8 kΩ fixed resistor, 10 kΩ fixed resistor, 15 kΩ fixed resistor, 20 kΩ fixed resistor, 22 kΩ fixed resistor, 68 kΩ fixed resistor, and 82 kΩ fixed resistor.
- SPST switch (3) or SPDT switch (3).

GLOSSARY OF TERMS

Operational amplifier (op amp) A high-gain amplifier that can increase AC or DC voltages

741 op amp The standard eight-pin operational amplifier IC used today in most commercial applications

Offset null terminals Two terminals or pins on an op amp that are used to adjust or "calibrate" the output of an op amp to 0 volts when it has no input voltage(s) applied; they are used only for applications involving sensitive measurements

Inverting input terminal One of the input pins of an op amp, represented on a schematic symbol with a negative sign (–). Any AC signal applied to this input terminal will undergo a 180° phase shift by the time it reaches

the output terminal; that is, the signal will be inverted, or "flipped over" as well as amplified

Non-inverting input terminal One of the input pins of an op amp, represented on a schematic symbol with a positive sign (+). Any AC signal applied to this input will be amplified but unchanged in polarity by the time it reaches the output

Negative DC supply voltage terminal A terminal or pin of an op amp that is connected to a negative (–) DC supply voltage source, which supplies operating power to the op amp

Positive DC supply voltage terminal A terminal or pin of an op amp that is connected to a positive (+) DC supply voltage source, which supplies operating power to the op amp

Open-loop voltage gain A_{OL} The gain of an op amp without a feedback circuit

Comparator An op amp configuration that "compares" applied input voltage(s) and produces a resulting output voltage

Voltage follower An op amp configuration that has a wire connected from the output terminal to the inverting terminal input, limiting the op amp's A_V to 1. The voltage follower, like the BJT emitter follower and the FET source follower, is used only to match a high impedance input stage to a low impedance output stage

Non-inverting amplifier An op amp configuration that increases an AC voltage applied to its inverting input terminal; it uses resistors that provide negative

feedback to control the op amp voltage gain

Feedback resistor R_f A resistor connected between the output terminal and the inverting terminal of an op amp that is used to provide negative feedback, thus controlling the gain of an op amp

Input resistor R_i A resistor connected to an input terminal of an op amp that works with the feedback resistor R_f to control the gain of an op amp

Closed-loop voltage gain A_{CL} The gain of an op amp with a feedback circuit that contains a feedback resistor R_f and an input resistor R_i

Inverting amplifier An op amp configuration that increases and inverts an AC voltage applied to its inverting input terminal; it uses resistors that provide negative feedback to control the op amp voltage gain

Differential amplifier An op amp configuration that provides an amplified output voltage based on the difference of two input voltages

Summing amplifier An op amp configuration that provides an amplified output voltage based on the sum of two or more input voltages

Frequency compensation capacitor A capacitor in an op amp IC used to limit the high-frequency operation of the op amp

8-1 INTRODUCTION TO OPERATIONAL AMPLIFIERS (OP AMPS)

The **operational amplifier (op amp)** is a high-gain differential amplifier that can amplify AC or DC voltages. The amplifiers we discussed in previous chapters used a DC voltage (bias) to supply operating power for the transistors, but the transistors themselves amplified *only* AC voltages. The op amp also uses DC input power, but it can amplify both AC input *and* DC input voltages. To avoid confusion and maintain consistency with previous chapters, we'll use the op amp as an AC amplifier until we reach the section on summing amplifiers.

The op amp has high input impedance and low output impedance. An op amp's input impedance is typically @ 300 kΩ to 2 MΩ, and its output impedance is @ 100 Ω. The open-loop voltage gain of an op amp (no feedback circuitry) can vary from 100,000 to 1 million. Early op amps were used in analog computers for mathematical operations such as addition and subtraction. That's where the name *operational* amplifier originated.

Karl D. Swartzel, Jr., invented the op amp in 1941 at Bell Labs. The first op amp was made of vacuum tubes and was used in World War II as part of a system to direct artillery rounds. By 1961, solid-state op amps made of multiple discrete (individual) parts appeared on the market. In 1963, Bob Widlar at Fairchild Semiconductor designed the first op amp IC. Five years of refinement led to the **741 op amp**, the standard op amp IC in use today.

Figure 8-1 shows the *internal* schematic of a 741 op amp. Before 1968, an op amp circuit would look like Figure 8-1, having many discrete (individual) parts connected together.

The circuit of Figure 8-1 has 35 electronic parts—transistors, resistors, one diode, and one capacitor—that would have taken up as much space as a slice of bread. Today, these parts are manufactured onto one IC, or chip, which is about the size of a dime. Figure 8-2 shows two common 8-pin 741 ICs used today. The IC on the left with a single op amp is the LM741CN, and the one on the right is the LM1458N, which contains two op amps in one IC.

The single op amp IC LM741CN in Figure 8-2 has the NTE replacement number 941M. Figure 8-3(a) shows two spec sheets for the NTE replacement numbers 941M and 941SM.

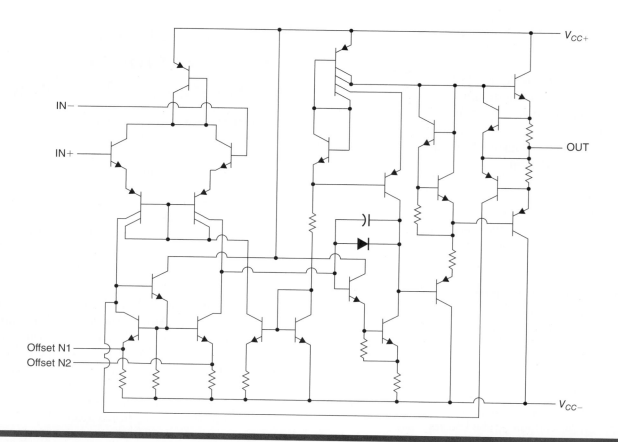

FIGURE 8-1   Internal schematic of a 741 op amp (Courtesy of Texas Instruments Incorporated)

The pin connection diagram for the op amp is listed on the second sheet in Figure 8-3(b). We discussed the op amp briefly in Chapter 3, and now we'll go into greater detail.

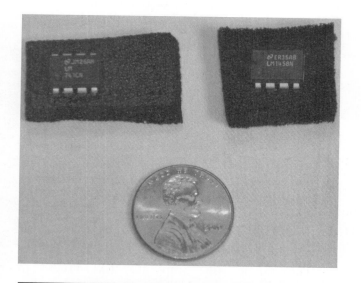

FIGURE 8-2   LM741CN and LM1458N ICs (Courtesy of Tracy Grace Leleux)

Although the internal circuitry of an op amp contains transistors, resistors, one diode, and one capacitor, the IC is a sealed unit. Thus, electronic technicians have access to only the eight external pins, and a technician has to know the pin numbers of an IC to work with it. If one of the internal transistors or resistors burns out, the entire IC must be replaced.

Looking at the spec sheet of Figure 8-3(b), notice the notch on the top of the IC in the pin connection diagram. Remember, manufacturers of ICs often put a dot or a notch on an IC for identification purposes. When building a circuit with any IC, if the notch is up or at the 12 o'clock position, pin 1 is the first pin on the left. The remaining pins are numbered in counter-clockwise sequence.

If there is a dot on an IC, then the pin closest to the dot is pin 1. Both op amp ICs of Figure 8-2 are unusual in that they each have a dot *and* a notch.

Pins 1 and 5 are called the **offset null terminals**, or pins. These pins are used to adjust or calibrate the output of an op amp to 0 volts when it has no input voltage(s) applied. The offset

ELECTRONICS, INC.
44 FARRAND STREET
BLOOMFIELD, NJ 07003
(973) 748–5089
http://www.nteinc.com

NTE941M & NTE941SM
Integrated Circuit
Operational Amplifier

Description:

The NTE941M and NTE941SM (Surface Mount) are general purpose operational amplifiers in 8–Lead DIP type packages and offer many features which make their application nearly foolproof: overload protection on the input and output, no latch–up when the common mode range is exceeded, as well as freedom from oscillators.

Absolute Maximum Ratings:

Supply Voltage, V_S . . . . ±18V

Differential Input Voltage, V_{ID} . . . . ±30V

Common Mode Input Voltage (Note 2), V_{ICM} . . . . ±15V

Power Dissipation (Note 1), P_D 500mW

Output Short–Circuit Duration, t_S . . . . Continuous

Operating Temperature Range, T_{opr} . . . . 0° to +70°C

Storage Temperature Range, T_{stg} . . . . −65° to +150°C

Junction Temperature, T_J . . . . +100°C

Lead Temperature (During Soldering, 10sec), T_L . . . . +260°C

Thermal Resistance, Junction–to–Ambient, R_{thJA}

    NTE941M . . . . +100°C/W

    NTE941SM . . . . +195°C/W

Note 1. For operation at elevated temperatures, these devices must be derated based on thermal resistance, and T_J Max ($T_J = T_A + (R_{thJA} P_D)$).

Note 2. For supply voltage less than ±15V, the absolute maximum input voltage is equal to the supply voltage.

FIGURE 8-3(a) Specification sheets for op amp LM741CN, NTE 941M (Courtesy of NTE Electronics, Inc.)

Electrical Characteristics: ($V_S = \pm 15V$, $0° \leq T_A \leq +70°C$ unless otherwise specified)

Parameter	Symbol	Test Conditions		Min	Typ	Max	Unit
Input Offset Voltage	V_{IO}	$R_S \leq 10k\Omega$	$T_A = +25°C$	–	2.0	6.0	mV
				–	–	7.5	mV
Input Offset Voltage Adjustment Range	V_{IOR}	$V_S = \pm 20V$, $T_A = +25°C$		–	± 15	–	V
Input Offset Current	I_{IO}	$T_A = +25°C$		–	20	200	nA
				–	–	300	nA
Input Bias Current	I_{IB}	$T_A = +25°C$		–	80	500	nA
				–	–	0.8	μA
Input Resistance	r_i	$V_S = \pm 20V$, $T_A = +25°C$		0.3	2.0	–	$M\Omega$
Common Mode Input Voltage Range	V_{ICR}	$T_A = +25°C$		–	± 12	± 13	V
Large Signal Voltage Gain	A_V	$V_O = \pm 10V$, $R_L \geq 2k\Omega$	$T_A = +25°C$	20	200	–	V/mV
				15	–	–	V/mV
Output Voltage Swing	V_O	$R_L \geq 10k\Omega$		± 12	± 14	–	V
		$R_L \geq 2k\Omega$		± 10	± 13	–	V
Output Short–Circuit Current	I_{OS}	$T_A = +25°C$		–	25	–	mA
Common–Mode Rejection Ratio	CMRR	$R_S \leq 10k\Omega$, $V_{CM} = \pm 12V$		70	90	–	dB
Supply Voltage Rejection Ratio	PSRR	$V_S = \pm 20V$ to $\pm 5V$, $R_S \leq 10k\Omega$		77	96	–	dB
Transient Response Rise Time	t_{TLH}	$T_A = +25°C$, Unity Gain		–	0.3	–	μs
Transient Response Overshoot	os			–	5	–	%
Transient Response Slew Rate	SR			–	0.5	–	V/μs
Supplu Current	I_D	$T_A = +25°C$		–	1.7	2.8	mA
Power Consumption	P_C	$T_A = +25°C$		–	50	85	mW

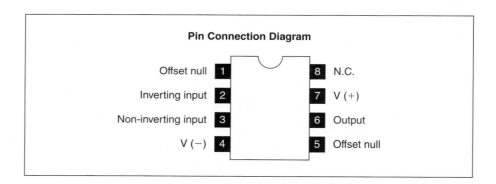

Pin Connection Diagram

Offset null	1	8	N.C.
Inverting input	2	7	V (+)
Non-inverting input	3	6	Output
V (−)	4	5	Offset null

FIGURE 8-3(b) (continued)

null pins are used only for applications involving sensitive measurements, such as the instruments used in offshore oil and gas drilling. Most op amp circuits don't use the offset null pins, but we'll still discuss them later in this chapter.

Pin 2 is called the **inverting input terminal**; it is represented on a schematic symbol with a negative sign (−). Any AC signal applied to this input terminal will undergo a 180° phase shift by the time it reaches the output terminal at pin 6; that is, the signal will be inverted or flipped over as well as amplified.

Pin 3 is the **non-inverting input terminal**. Any AC signal applied to this input will be amplified but unchanged in polarity by the time it reaches the output at pin 6.

Pin 4 is the **negative DC supply voltage terminal**. Like the transistor amplifiers we discussed in previous chapters, op amps amplify an AC signal, but they need a DC supply voltage. The op amp is unusual in electronics because it can be powered by a negative DC supply voltage, a positive DC supply voltage, or both. Recall that a negative DC supply voltage is not the same as ground. A negative DC voltage is "below ground"; that is, it is tapped from the negative alternation of a rectified AC voltage source. Most power supplies for electronic labs provide both positive and negative voltages.

Pin 6 is the output terminal. An op amp has only one output terminal.

Pin 7 is the **positive DC supply voltage terminal**. Pin 7 and pin 4 supply an op amp with DC voltages, also called the rail voltages. An op amp's AC output signal voltage (V_{RMS}, or DC equivalent) can't exceed the rail voltage(s) because the AC signal steals power from the DC supply voltage to get larger. In fact, for any op amp circuit, the *maximum* AC output signal voltage (V_{RMS}) will typically be one or two volts lower than the rail voltage. This is because the internal circuitry of an op amp IC also uses some of the DC supply voltage—the internal transistors and resistors have small DC voltages dropped across them.

Link to Prior Learning

In BJT and FET amplifiers, the input AC signal voltage gets larger by stealing power from the DC supply voltage. The same holds true for an op amp.

Pin 8 is labeled "NC," for no connection. This pin has no internal connection to the op amp. All ICs are manufactured with an even number of pins so that the IC won't tilt to one side when it's mounted on a circuit board. Pin 8 simply provides balance.

Remember, you can't tell that an IC is an op amp just by looking at it. The only way to tell the pin configuration of an op amp is by looking up its part number in the *NTE* catalog or another semiconductor replacement guide. *Look it up to hook it up.*

Before we discuss the operation of an op amp, let's look at the schematic symbols used for op amps. Figure 8-4 shows three schematic symbols commonly used for op amps.

The standard schematic symbol for an op amp is shown in Figure 8-4(a). This symbol has only the inverting input, non-inverting input, and output terminals labeled. This is also called a three-terminal op amp.

Figure 8-4(b) shows the positive and negative DC supply voltage terminals added. This is also called a five-terminal op amp.

Figure 8-4(c) has the offset null terminals, or pins, also included. This is called a seven-terminal op amp.

No matter which of the three op amp schematic symbols are shown in a circuit diagram, an op amp has seven possible working terminals and one terminal that has no internal connection.

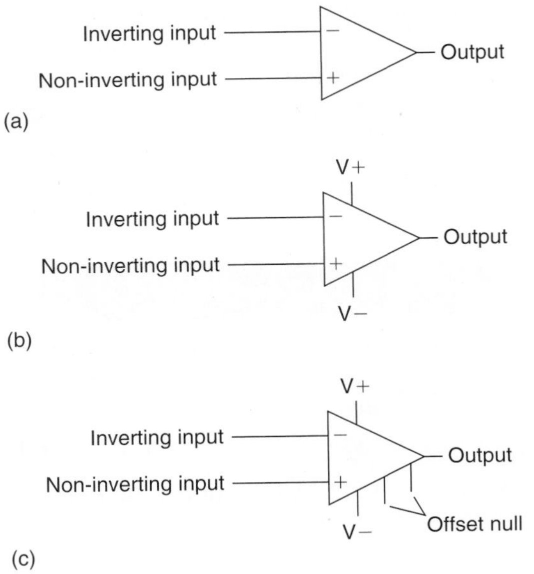

FIGURE 8-4 Three schematic symbols used for op amps: a) Standard symbol, b) DC power supply inputs added, and c) Offset null pins added (© Cengage Learning 2012)

Although op amps are used in hundreds of applications, there are six basic op amp configurations: comparator, voltage follower, differential amplifier, non-inverting amplifier, inverting amplifier, and summing amplifier. The comparator configuration uses no external feedback circuitry. The voltage follower, non-inverting amplifier, inverting amplifier, differential amplifier, and summing amplifier all use some type of external feedback circuitry. Because an op amp performs differently in each configuration, we'll approach the operation of an op amp by looking at each of these configurations.

8-2 OP AMP COMPARATOR

The op amp has an **open-loop voltage gain** that can vary from 100,000 to 1 million. Open-loop voltage gain means there is no feedback circuitry; it is written as A_{OL}. An op amp without feedback circuitry really acts as a **comparator** because it compares the input voltage(s) and produces a resulting output voltage. The open-loop voltage gain can vary significantly from op amp to op amp. Also, there's no way to control the gain without additional circuitry. Figure 8-5 shows an op amp configured for an open-loop gain.

Recall that inside the op amp IC are transistors, resistors, one diode, and one capacitor working together to produce an output voltage. The op amp is a differential amplifier: its output voltage is an amplified version of the difference between the voltages on its input terminals. Put simply, the V_{IN} of an open-loop op amp is the voltage at the non-inverting input minus the voltage at the inverting input.

EQUATION 8-1

Comparator open-loop input voltage

$$V_{IN} = V_{NONINV} - V_{INV}$$

V_{NONINV} = voltage at non-inverting terminal

V_{INV} = voltage at inverting terminal

Let's apply this equation to the circuit of Figure 8-5. Notice that the DC supply, or rail, voltages are $+15$ V and -15 V; this information will prove important later. For Figure 8-5, the inverting input is grounded, so its voltage is considered 0 V.

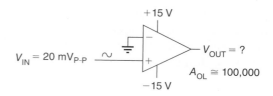

FIGURE 8-5 Op amp configured for open-loop gain (© Cengage Learning 2012)

So, for Figure 8-5,

$$V_{IN} = V_{NONINV} - V_{INV} = 20 \text{ mV}_{P-P} - 0 \text{ V}$$
$$= 20 \text{ mV}_{P-P}$$

Now, let's bring forward from Chapter 5 the generic equation for an amplifier voltage gain.

$$A_V = \frac{V_{OUT}}{V_{IN}}$$

First, we'll change A_V to A_{OL} since this is an open-loop voltage gain configuration. If we rewrite the equation, we can solve for V_{OUT}.

$$V_{OUT} = (V_{IN})(A_{OL})$$

We mentioned earlier that an op amp has an open-loop voltage gain A_{OL} of @ 100,000 to one million. Let's use the lower value of 100,000.

So, for Figure 8-5,

$$V_{OUT} = (V_{IN})(A_{OL}) = (20 \text{ mV}_{P-P})(100,000)$$
$$= 2,000 \text{ V}_{P-P}$$

This value sounds incredibly high, and it is. Remember, an op amp's AC output signal voltage (V_{RMS}) can't exceed the rail voltage(s) because the AC signal steals power from the DC supply voltage to get larger. For any op amp circuit, the *maximum* AC output signal voltage (V_{RMS}) will typically be one or two volts lower than the rail voltage(s). So, for the circuit of Figure 8-5, the *actual* AC output voltage will be @ 13 V_{RMS} to 14 V_{RMS} since the rail voltages are $+15$ V and -15 V. This AC output voltage will also be distorted because the op amp has gone beyond saturation. Remember, driving an amplifier beyond saturation can result in clipping of the waveform. Thus, the open-loop comparator circuit of Figure 8-5 is impractical without additional circuitry. However, it does demonstrate the basic operation of an op amp. Let's do another example to reinforce the calculations we've just learned.

EXAMPLE 1

Situation

Determine V_{IN} and V_{OUT} for the circuit of Figure 8-6. Use 150,000 for the A_{OL}.

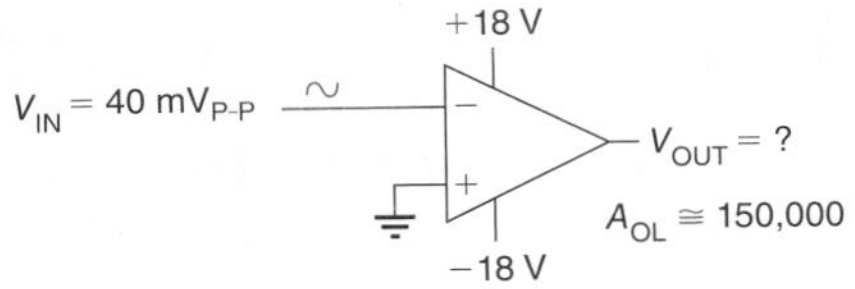

FIGURE 8-6   Op amp configured for open-loop gain (© Cengage Learning 2012)

Solution

For Figure 8-6, the non-inverting input is grounded, so its voltage is considered 0 V.

So, for Figure 8-6,

$$V_{IN} = V_{NONINV} - V_{INV} = 0\ V - 40\ mV_{P\text{-}P}$$
$$= -40\ mV_{P\text{-}P}$$

Then, $V_{OUT} = (V_{IN})(A_{OL})$
$$= (-40\ mV_{P\text{-}P})(150,000)$$
$$= -6,000\ mV_{P\text{-}P}$$

EXAMPLE 2

Situation

Determine V_{IN} and V_{OUT} for the circuit of Figure 8-7. Use 200,000 for the A_{OL}.

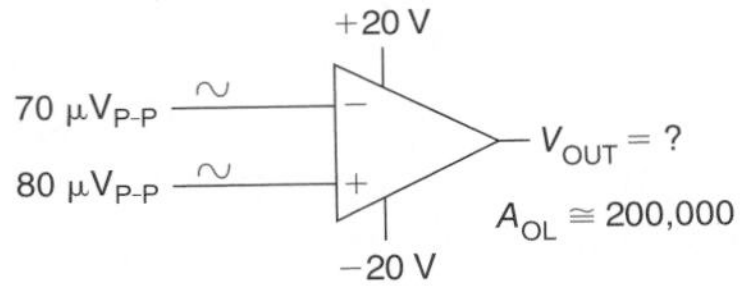

FIGURE 8-7   Op amp configured for open-loop gain (© Cengage Learning 2012)

Solution

For Figure 8-7, different voltages are applied to each input.

So, for Figure 8-7,

$$V_{IN} = V_{NONINV} - V_{INV} = 80\ \mu V_{P\text{-}P} - 70\ \mu V_{P\text{-}P}$$
$$= 10\ \mu V_{P\text{-}P}$$

Then, $V_{OUT} = (V_{IN})(A_{OL})$
$$= (10\ \mu V_{P\text{-}P})(200,000)$$
$$= 2\ V_{P\text{-}P}$$

Again, this value sounds incredibly high, and it is. For the circuit of Figure 8-6, the *actual* AC output voltage will be @ $-16\ V_{RMS}$ to $-17\ V_{RMS}$ since the rail voltages are $+18\ V$ and $-18\ V$. This AC output voltage will be inverted since the input voltage was applied to the inverting terminal and distorted; thus, the open-loop comparator circuit of Figure 8-6 is impractical without additional circuitry. The output voltage is negative because the input voltage is applied to the inverting terminal.

For the circuit of Figure 8-7, the actual AC output voltage will be @ $2\ V_{P\text{-}P}$ with little or no distortion because it's within the rail voltage limits of $+20\ V$ and $-20\ V$. What will be the *measured* AC output voltage (RMS) displayed on a DMM?

Link to Prior Learning

The peak voltage in an AC circuit equals the peak-to-peak voltage divided by two.

$$V_P = \frac{V_{P\text{-}P}}{2}$$

The RMS voltage (DC equivalent) in any AC circuit equals 0.707 multiplied by the peak voltage.

$$V_{RMS} = 0.707(V_P)$$

Using these two AC equations, the measured V_{OUT} for the circuit of Figure 8-7 will be 0.707 V.

$$V_P = \frac{V_{P-P}}{2} = \frac{2\ V}{2} = 1\ V$$

$$V_{RMS} = 0.707(V_P) = 0.707(1\ V) = 0.707\ V$$

Knowing how to calculate the measured AC output voltage will come in handy later when we do lab activity. In Figure 8-7, the 200,000 open-loop gain is much higher than we've seen for the BJT and FET amplifiers that we've studied so far.

The results of these circuits have demonstrated that the open-loop voltage gain of an op amp can be extremely high—in some cases too high, driving the amplifier beyond saturation and causing distortion of the AC output signal. Unless the AC input voltage is in the μV range, the open-loop configuration is of little use. For this reason, op amps usually use additional circuitry involving feedback to control the gain.

Now, we'll look at the op amp configurations that use feedback circuitry. We will start with the simplest circuit, the voltage follower.

8-3 OP AMP VOLTAGE FOLLOWER

The **voltage follower** configuration takes advantage of an op amp's input and output impedances. An op amp's input impedance is typically @ 300 kΩ to 2 MΩ and its output impedance @ 100 Ω. Figure 8-8 shows the op amp voltage follower configuration.

Notice in the circuit of Figure 8-8 that a wire is connected from the output terminal to the inverting input terminal. This means that the A_V of the op amp voltage follower is 1, so there really is no gain because anything multiplied by 1 equals itself. Thus, like the BJT emitter follower and the FET source follower, the voltage follower can be used only to match a high impedance input stage to a low impedance output stage. When used to match

impedances, a voltage follower is often called a buffer. The voltage follower is used in radio receivers as a buffer for the local oscillator, and it's also used in amplifiers to match the output impedance of an audio amplifier to one or more loudspeakers.

8-4 OP AMP NON-INVERTING AMPLIFIER

The op amp **non-inverting amplifier** uses resistors that provide negative feedback to control the op amp voltage gain. Figure 8-9 shows an op amp configured as a non-inverting amplifier.

Notice in the circuit of Figure 8-9 that a 10 kΩ resistor R_f has been connected between the output terminal and the inverting terminal. This resistor is called the **feedback resistor**, and it provides negative feedback that will reduce the gain to a manageable level. *Yes, negative feedback reduces gain.* Remember, an op amp without feedback circuitry can provide too much gain, resulting in distortion of the AC input signal.

The 2 kΩ resistor R_i is an **input resistor** that works with R_f to control the gain. In fact, resistors R_f and R_i determine the gain. Notice that R_i is also grounded, and the AC input signal enters at the non-inverting terminal. Thus, the key visual features for a non-inverting op amp are a feedback resistor, a grounded input resistor, and an AC signal applied to the non-inverting terminal.

Technicians can change the value of R_f or R_i or both and thus change the gain of an op amp. This is a major advantage of the op amp amplifier when compared to BJT and FET amplifiers. With an op amp, you only have to change either R_f or R_i to change the op amp's gain. Changing the gain of BJT and FET amplifiers is more complicated.

When an op amp uses feedback circuitry, the gain is called the **closed-loop voltage gain (A_{CL})**.

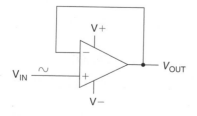

FIGURE 8-8 Op amp voltage follower
(© Cengage Learning 2012)

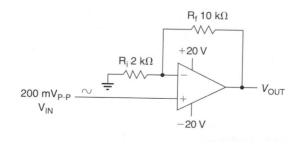

FIGURE 8-9 Op amp non-inverting amplifier
(© Cengage Learning 2012)

To determine the static voltage gain (no AC signal applied) for a non-inverting op amp, use Equation 8-2.

EQUATION 8-2

Non-inverting op amp static voltage gain

$$A_V = \frac{R_f}{R_i} + 1$$

R_f = feedback resistance

R_i = input resistance

For the circuit of Figure 8-9, the AC output voltage should be @ 1.2 V_{P-P} with little or no distortion because it's within the rail voltage limits of + 20 V and –20 V. Let's do another.

For the circuit of Figure 8-10, the AC output voltage should be @ 3.95 V_{P-P} with little or no distortion because it's within the rail voltage limits of + 18 V and –18 V.

What would the V_{OUT} be if we measured it on our DMM, which measures RMS voltages?

$$V_P = \frac{V_{P-P}}{2} = \frac{3.95\ V}{2} = 1.975\ V$$

$$V_{RMS} = 0.707(V_P) = 0.707(1.975\ V) = 1.39\ V$$

EXAMPLE 3

Situation

Determine the static voltage gain A_V and V_{OUT} for the circuit of Figure 8-9.

Solution

For Figure 8-9,

$$A_V = \frac{R_f}{R_i} + 1 = \frac{10\ k\Omega}{2\ k\Omega} + 1 = 5 + 1 = 6$$

Then, $V_{OUT} = (V_{IN})(A_V)$

$$= (200\ mV_{P-P})(6)$$

$$= 1.2\ V_{P-P}$$

Thus, the V_{OUT} for the circuit of Figure 8-10 will show @ 3.95 V_{P-P} on an o-scope and @ 1.39 V on a DMM.

Applications involving the non-inverting amplifier include video amplifiers that connect to typical 75 Ω cable televisions and low-voltage display screens used on cell phones. Let's build a live version of the non-inverting op amp.

EXAMPLE 4

Situation

Determine the static voltage gain A_V and V_{OUT} for the circuit of Figure 8-10.

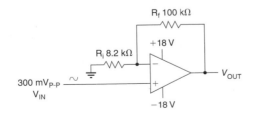

FIGURE 8-10 Op amp non-inverting amplifier (© Cengage Learning 2012)

Solution

For Figure 8-10,

$$A_V = \frac{R_f}{R_i} + 1 = \frac{100\ k\Omega}{8.2\ k\Omega} + 1$$

$$= 12.19 + 1$$

$$= 13.19$$

Then, $V_{OUT} = (V_{IN})(A_V)$

$$= (300\ mV_{P-P})(13.19)$$

$$= 3.95\ V_{P-P}$$

LAB ACTIVITY 8-1

Op Amp Non-Inverting Amplifier

Materials, Equipment, and Parts:

- *NTE* catalog or Internet access, www.nteinc. com

- PC w/Multisim®, Electronics Workbench®, or SPICE.

- DMM with test leads.

- Dual trace oscilloscope w/BNC-to-alligator leads.

- Positive DC voltage source, 0–20 V.

- Negative DC voltage source, 0–20 V.

- Function generator, 1 kHz to 10 kHz.

- Breadboard and connecting wires.

- LM741CN, UA741CN, or equivalent op amp IC.

- 1 kΩ fixed resistor (2), 3.9 kΩ fixed resistor, 10 kΩ fixed resistor, 22 kΩ fixed resistor, and 82 kΩ fixed resistor.

Discussion Summary:

The op amp non-inverting amplifier takes an AC voltage applied to its non-inverting input terminal and increases it. The voltage gain A_V is determined by the feedback resistor R_f and the input resistor R_i. The maximum AC output voltage (RMS) is determined by the DC supply voltages, or rail voltages.

Procedure:

SAFETY FIRST. Eye protection should always be worn when working with live voltages. Before powering on a live circuit, always check with your instructor.

1 Write the part number of the op amp at your workstation. Then use the *NTE* catalog or visit the website www.nteinc.com to find and record the NTE replacement number, diagram number, and V_{CC}.

Part number _____

NTE number _____

Diagram # _____

V_{CC} _____

2 Draw the op amp pin configuration diagram and label each pin.

(continues)

LAB ACTIVITY 8-1

(continued)

3 Build the circuit of Figure 8-11 on your breadboard.

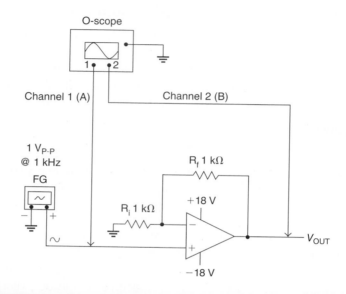

FIGURE 8-11   Op amp non-inverting amplifier (© Cengage Learning 2012)

4 Connect the positive lead of the +18 V DC voltage source to pin 7 of the op amp.

5 Connect the positive lead of the −18 V DC voltage source to pin 4 of the op amp.

6 Connect the positive lead of the function generator to the non-inverting terminal of the op amp, pin 3.

7 Connect the positive lead of the oscilloscope's Channel 1 to the non-inverting terminal of the op amp, pin 3.

8 Connect the positive lead of the o-scope's Channel 2 to the output terminal of the op amp, pin 6.

9 Connect the black lead of the function generator, the black lead of the +18 V DC voltage source, the black lead of the −18 V DC voltage source, the black leads of the o-scope, and a wire from the input resistor R_i to one point (hole) on the breadboard, which is ground.

10 Set the function generator to the sine wave setting and to an amplitude of 1 V peak-to-peak. You may have to use your o-scope's Channel 1 volts/division setting to adjust the voltage to 1 V_{P-P}.

LAB ACTIVITY 8-1

11 Set the function generator at a frequency of 1 kHz.

12 Have the instructor check your circuit.

13 Calculate and record the *static* A_V of the op amp non-inverting amplifier.

$$A_V = \frac{R_f}{R_i} + 1 =$$

Static A_V _____

14 Power on the +18 V DC voltage source, the −18 V DC voltage source, the function generator, and the o-scope. Use your o-scope to ensure the AC input signal on Channel 1 is 1 V peak-to-peak. You may have to adjust the Channel 1 volts/division setting and/or other o-scope settings. Record the value below.

V_{IN} _____

15 Measure and record the AC output voltage that appears on the o-scope's Channel 2. You may have to adjust the Channel 2 volts/division setting and/or other o-scope settings. Record the value below.

V_{OUT} _____

16 Use the results of Steps 14 and 15 and the equation below to calculate and record the dynamic voltage gain of the op amp non-inverting amplifier.

$$A_V = \frac{V_{OUT}}{V_{IN}} =$$

Dynamic A_V _____

How does this dynamic (measured) A_V compare to the calculated static A_V of Step 13? Explain any differences.

17 Set your o-scope's vertical mode switch to the dual setting. Adjust your o-scope settings so that the 1 V_{P-P} input signal is on top and the V_{OUT} signal is directly below it. Are the waveforms in phase (peaking at the same time) or out of phase?

18 Power off the +18 V DC voltage source, the −18 V DC voltage source, the function generator, and the o-scope. Replace R_f with a 3.9 kΩ resistor.

(continues)

LAB ACTIVITY 8 - 1

(continued)

19 Recalculate and then record the static gain.

$$A_V = \frac{R_f}{R_i} + 1 =$$

Static A_V _____

Power on the +18 V DC voltage source, the −18 V DC voltage source, the function generator, and the o-scope. Use your o-scope to ensure the AC input signal on Channel 1 is 1 V peak-to-peak. Record the value below.

V_{IN} _____

Measure and record the AC output voltage that appears on the o-scope's Channel 2. You may have to adjust the Channel 2 volts/division setting and/or other o-scope settings. Record the value below.

V_{OUT} _____

Use these values to calculate and record the dynamic voltage gain of the op amp non-inverting amplifier.

$$A_V = \frac{V_{OUT}}{V_{IN}} =$$

Dynamic A_V _____

How does this dynamic (measured) A_V compare to the calculated static A_V? Explain any differences. Power off.

20 Replace R_f with a 10 kΩ resistor and then repeat Step 19. Record the values below.

$$A_V = \frac{R_f}{R_i} + 1 =$$

Static A_V _____

V_{IN} _____

V_{OUT} _____

$$A_V = \frac{V_{OUT}}{V_{IN}} =$$

Dynamic A_V _____

How does this dynamic (measured) A_V compare to the calculated static A_V? Explain any differences. Power off.

LAB ACTIVITY 8-1

 21 Replace R_f with a 22 kΩ resistor and then repeat Step 19. Record the values below.

$$A_V = \frac{R_f}{R_i} + 1 =$$

Static A_V _____

V_{IN} _____

V_{OUT} _____

$$A_V = \frac{V_{OUT}}{V_{IN}} =$$

Dynamic A_V _____

How does this dynamic (measured) A_V compare to the calculated static A_V? Explain any differences. Power off.

 22 Replace R_f with an 82 kΩ resistor and then repeat Step 19. Record the values below.

$$A_V = \frac{R_f}{R_i} + 1 =$$

Static A_V _____

V_{IN} _____

When you power on this time, notice that the V_{OUT} is distorted—the positive and negative peaks of the AC output waveform are cut off or clipped. This AC output voltage should be @ 83 V_{P-P} (@ 29.34 V_{RMS}), but it's hard to get an accurate o-scope reading because of the distortion. Set your DMM to measure AC volts on a setting higher than 20 V. Place the positive lead of the DMM to the output terminal of the op amp, pin 6, and connect the negative lead of the DMM to ground. Record this RMS voltage.

V_{OUT} _____ (RMS)

Now, convert this RMS voltage to V_{P-P}:

$$V_P = \frac{V_{RMS}}{0.707} =$$

$$V_{P-P} = 2(V_P) =$$

(continues)

LAB ACTIVITY 8-1

(continued)

Now, plug this value into the V_{OUT} slot of the A_V equation.

$$A_V = \frac{V_{OUT}}{V_{IN}} =$$

Dynamic A_V _____

How does this dynamic (measured) A_V compare to the calculated static A_V? Explain any differences. (Hint: Think rail voltage.) Power off.

 Have the instructor verify your results.

 Build the circuit on Multisim®, Electronics Workbench®, or SPICE and then compare the results.

8-5 Op Amp Inverting Amplifier

The op amp **inverting amplifier** also uses external resistors to provide negative feedback and control (reduce) the op amp voltage gain. An op amp inverting amplifier both inverts and amplifies an AC signal; that is, the signal will be flipped over as well as amplified. Remember, the BJT common-emitter amplifier and the FET common-source amplifier also produced a 180° phase shift of the AC input signal. Figure 8-12 shows an op amp configured as an inverting amplifier.

Notice in the circuit of Figure 8-12 that a 47 kΩ resistor R_f has been connected between the output terminal and the inverting terminal. Again, this resistor is called the feedback resistor, and it provides negative feedback that will reduce the gain to a manageable level.

The 6.8 kΩ resistor R_i is an input resistor that works with R_f to control the gain. In fact, resistors R_f and R_i determine the gain. Notice that the non-inverting terminal is grounded, and the AC input signal enters at the inverting terminal. Thus, the key visual features for an inverting op amp are a feedback resistor, an input resistor, a grounded non-inverting terminal, and an AC signal applied to the inverting terminal.

Internet Alert

Check the website http://www.falstad.com/circuit/e-amp-invert.html to see the operation of an op amp inverting amplifier.

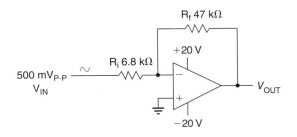

FIGURE 8-12 Op amp inverting amplifier
(© Cengage Learning 2012)

Technicians can change the value of R_f or R_i or both and thus change the gain of an op amp. When an op amp uses feedback circuitry, the gain is called the closed-loop gain, or A_{CL}. To determine the static voltage gain (no AC signal applied) for an inverting op amp, use Equation 8-3.

EQUATION 8-3

Inverting op amp static voltage gain

$$A_V = \frac{R_f}{R_i}$$

R_f = feedback resistance

R_i = input resistance

EXAMPLE 5

Situation

Determine the static voltage gain A_V and V_{OUT} for the circuit of Figure 8-12.

Solution

For Figure 8-12,

$$A_V = \frac{R_f}{R_i} = \frac{47 \text{ k}\Omega}{6.8 \text{ k}\Omega} = 6.91$$

$$\text{Then, } V_{OUT} = (V_{IN})(A_V)$$

$$= (500 \text{ mV}_{P-P})(6.91)$$

$$= -3.45 \text{ V}_{P-P}$$

For the circuit of Figure 8-12, the AC output voltage should be @ –3.45 V_{P-P} with little or no distortion because it's within the rail voltage limits of +20 V and –20 V. Note that the V_{OUT} is a negative value because the AC input signal applied to the inverting terminal is positive.

Up to this point, we've talked of op amps as amplifying AC input signals; op amps can also work with DC input voltages. Recall that the internal transistors and resistors in an op amp have small DC voltages dropped across them. In most AC applications, these voltage drops are minimal and don't affect the overall operation. However, these voltage drops can affect precise DC measurements, so

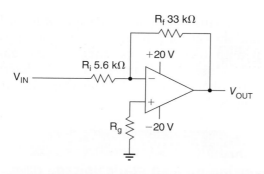

FIGURE 8-13 Op amp inverting amplifier with grounding resistor R_g (© Cengage Learning 2012)

sometimes a grounding resistor R_g is inserted between the non-inverting terminal and ground to reduce the effects of errors caused by voltage drops across internal transistors and resistors. Figure 8-13 shows an op amp inverting amplifier with a grounding resistor R_g.

The value of this grounding resistor is determined by Equation 8-4, which is the same formula for resistors in parallel.

EQUATION 8-4

Inverting op amp grounding resistor

$$R_g = \frac{(R_f)(R_i)}{R_f + R_i}$$

R_f = feedback resistance

R_i = input resistance

EXAMPLE 6

Situation

Determine the value of the grounding resistor R_g for the circuit of Figure 8-13.

Solution

For Figure 8-13,

$$R_g = \frac{(R_f)(R_i)}{R_f + R_i} = \frac{(33\ k\Omega)(5.6\ k\Omega)}{33\ k\Omega + 5.6\ k\Omega}$$

$$= 4.78\ k\Omega$$

A 4.78 kΩ resistor will reduce the effects of errors caused by voltage drops across internal transistors and resistors. Again, R_g is needed only for applications requiring precise measurements.

The applications for an op amp inverting amplifier include photodiode and phototransistor amplifiers in solar cells and panels, temperature sensor amplifiers, and high-pass and low-pass filters. Now, let's build the inverting op amp on the breadboard.

LAB ACTIVITY 8-2

Op Amp Inverting Amplifier

Materials, Equipment, and Parts:

- *NTE* catalog or Internet access, www.nteinc.com

- PC w/Multisim®, Electronics Workbench®, or SPICE.

- DMM with test leads.

- Dual trace oscilloscope w/BNC-to-alligator leads.

- Positive DC voltage source, 0–20 V.

- Negative DC voltage source, 0–20 V.

- Function generator, 1 kHz to 10 kHz.

- Breadboard and connecting wires.

- LM741CN, UA741CN, or equivalent op amp IC.

- 1 kΩ fixed resistor (2), 2.7 kΩ fixed resistor, 6.8 kΩ fixed resistor, 15 kΩ fixed resistor, and 68 kΩ fixed resistor.

Discussion Summary:

The op amp inverting amplifier increases and inverts an AC voltage applied to its inverting input terminal. The voltage gain A_V is determined by the feedback resistor R_f and the input resistor R_i. The maximum AC output voltage (RMS) is determined by the DC supply voltages, or rail voltages.

Procedure:

SAFETY FIRST. Eye protection should always be worn when working with live voltages. Before powering on a live circuit, always check with your instructor.

1 Write the part number of the op amp at your workstation. Then use the *NTE* catalog or visit the website www.nteinc.com to find and record the NTE replacement number, diagram number, and V_{CC}.

Part number _____

NTE number _____

Diagram # _____

V_{CC} _____

2 Draw the op amp pin configuration diagram and label each pin.

(continues)

LAB ACTIVITY 8 - 2

(continued)

3 Build the circuit of Figure 8-14 on your breadboard.

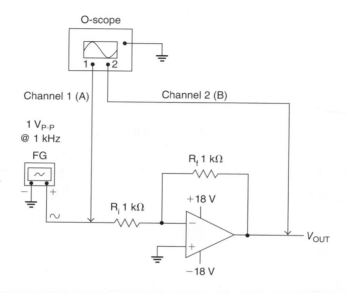

FIGURE 8-14   Op amp inverting amplifier (© Cengage Learning 2012)

4 Connect the positive lead of the +18 V DC voltage source to pin 7 of the op amp.

5 Connect the positive lead of the −18 V DC voltage source to pin 4 of the op amp.

6 Connect the positive lead of the function generator to the R_i, which connects to the inverting terminal of the op amp, pin 2.

7 Connect the positive lead of the o-scope's Channel 1 to the R_i, which connects to the inverting terminal of the op amp, pin 2.

8 Connect the positive lead of the o-scope's Channel 2 to the output terminal of the op amp, pin 6.

9 Connect the black lead of the function generator, the black lead of the +18 V DC voltage source, the black lead of the −18 V DC voltage source, the black leads of the o-scope, and a wire from the non-inverting terminal, pin 3, to one point (hole) on the breadboard, which is ground.

10 Set the function generator to the sine wave setting and to an amplitude of 1 V peak-to-peak. You may have to use your o-scope's Channel 1 volts/division setting to adjust the voltage to 1 V_{P-P}.

LAB ACTIVITY 8-2

11 Set the function generator at a frequency of 1 kHz.

12 Have the instructor check your circuit.

13 Calculate and record the *static* A_V of the op amp inverting amplifier.

$$A_V = \frac{R_f}{R_i} =$$

Static A_V _____

14 Power on the +18 V DC voltage source, the −18 V DC voltage source, the function generator, and the o-scope. Use your o-scope to ensure the AC input signal on Channel 1 is 1 V peak-to-peak. You may have to adjust the Channel 1 volts/division setting and/or other o-scope settings. Record the value below.

V_{IN} _____

15 Measure and record the AC output voltage that appears on the o-scope's Channel 2. You may have to adjust the Channel 2 volts/division setting and/or other o-scope settings. Record the value below.

V_{OUT} _____

16 Use the results of Steps 14 and 15 and the equation below to calculate and record the dynamic voltage gain of the op amp inverting amplifier.

$$A_V = \frac{V_{OUT}}{V_{IN}} =$$

Dynamic A_V _____

How does this dynamic (measured) A_V compare to the calculated static A_V of Step 13? Explain any differences.

17 Set your o-scope's vertical mode switch to the dual setting. Adjust your o-scope settings so that the 1 V_{P-P} input signal is on top and the V_{OUT} signal is directly below it. Are the waveforms in phase (peaking at the same time) or out of phase?

18 Power off the +18 V DC voltage source, the −18 V DC voltage source, the function generator, and the o-scope. Replace R_f with a 2.7 kΩ resistor.

(continues)

LAB ACTIVITY 8-2

(continued)

 Recalculate and then record the static gain.

$$A_V = \frac{R_f}{R_i} =$$

Static A_V _____

Power on the +18 V DC voltage source, the −18 V DC voltage source, the function generator, and the o-scope. Use your o-scope to ensure the AC input signal on Channel 1 is 1 V peak-to-peak. Record the value below.

V_{IN} _____

Measure and record the AC output voltage that appears on the o-scope's Channel 2. You may have to adjust the Channel 2 volts/division setting and/or other o-scope settings. Record the value below.

V_{OUT} _____

Use these values to calculate and record the dynamic voltage gain of the op amp inverting amplifier.

$$A_V = \frac{V_{OUT}}{V_{IN}} =$$

Dynamic A_V _____

How does this dynamic (measured) A_V compare to the calculated static A_V? Explain any differences. Power off.

 Replace R_f with a 6.8 kΩ resistor and then repeat Step 19. Record the values below.

$$A_V = \frac{R_f}{R_i} =$$

Static A_V _____

V_{IN} _____

V_{OUT} _____

$$A_V = \frac{V_{OUT}}{V_{IN}} =$$

Dynamic A_V _____

How does this dynamic (measured) A_V compare to the calculated static A_V? Explain any differences. Power off.

Replace R_f with a 15 kΩ resistor and then repeat Step 19. Record the values below.

$$A_V = \frac{R_f}{R_i} =$$

Static A_V _____

V_{IN} _____

V_{OUT} _____

$$A_V = \frac{V_{OUT}}{V_{IN}} =$$

Dynamic A_V _____

How does this dynamic (measured) A_V compare to the calculated static A_V? Explain any differences. Power off.

22 Replace R_f with a 68 kΩ resistor and then repeat Step 19. Record the values below.

$$A_V = \frac{R_f}{R_i} =$$

Static A_V _____

V_{IN} _____

When you power on this time, notice that the V_{OUT} is distorted: the positive and negative peaks of the AC output waveform are cut off or clipped. This AC output voltage should be @ 68 V_{P-P} (@ 24.03 V_{RMS}), but it's hard to get an accurate o-scope reading because of the distortion. Set your DMM to measure AC volts on a setting higher than 20 V. Place the positive lead of the DMM to the output terminal of the op amp, pin 6, and connect the negative lead of the DMM to ground. Record this RMS voltage.

V_{OUT} _____ (RMS)

Now, convert this RMS voltage to V_{P-P}:

$$V_P = \frac{V_{RMS}}{0.707} =$$

$$V_{P-P} = 2(VP) =$$

(continues)

LAB ACTIVITY 8-2

(continued)

Now, plug this value into the V_{OUT} slot of the A_V equation.

$$A_V = \frac{V_{OUT}}{V_{IN}} =$$

Dynamic A_V _____

How does this dynamic (measured) A_V compare to the calculated static A_V? Explain any differences. (Hint: Think of rail voltage.) Power off.

 Have the instructor verify your results.

 Build the circuit on Multisim®, Electronics Workbench®, or SPICE and then compare the results.

8-6 OP AMP DIFFERENTIAL AMPLIFIER

The op amp **differential amplifier** provides an amplified output voltage that is the difference of two input voltages. The differential amplifier works like the comparator except that it has external resistors to provide negative feedback and to control the voltage gain. Figure 8-15 shows an op amp configured as a differential amplifier.

Notice in the circuit of Figure 8-15 that there are four resistors: a feedback resistor R_f; a grounding resistor R_g; and a resistor at each input terminal. The R_f and R_g resistors are the same value, and the input resistors are the same value. Also, the differential amplifier has two input voltages.

The differential amplifier uses the same equation for V_{IN} as the comparator does, except that the overall input voltage is now called the difference voltage, or V_{DIFF}. The differential amplifier uses the same equation for voltage gain that the inverting amplifier does. The equations for the op amp differential amplifier are shown below.

EQUATION 8-5

Differential op amp input voltage difference

$$V_{DIFF} = V_{NONINV} - V_{INV}$$

V_{NONINV} = voltage at non-inverting terminal

V_{INV} = voltage at inverting terminal

EQUATION 8-6

Differential op amp static voltage gain

$$AV = \frac{R_f}{R_i}$$

R_f = feedback resistance

R_i = either input resistance

EQUATION 8-7

Differential op amp output voltage

$$V_{OUT} = (V_{DIFF})(A_V)$$

In Example 8-7, we'll use these equations to analyze the op amp differential amplifier.

For the circuit of Figure 8-15, the AC output voltage should be @ 114.89 mV$_{P-P}$ with little or no distortion because it's within the rail voltage limits of +20 V and −20 V. Let's do another.

EXAMPLE 7

Situation

Determine the input voltage difference, static voltage gain A_V, and V_{OUT} for the differential amplifier circuit of Figure 8-15.

Solution

For Figure 8-15,

$$V_{DIFF} = V_{NONINV} - V_{INV}$$

$$= 50\ mV - 30\ mV = 20\ mV$$

Then, $A_V = \dfrac{R_f}{R_i} = \dfrac{27\ k\Omega}{4.7\ k\Omega} = 5.74$

Then, $V_{OUT} = (V_{DIFF})(A_V)$

$$= (20\ mV_{P-P})(5.74)$$

$$= 114.89\ mV_{P-P}$$

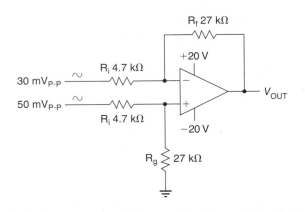

FIGURE 8-15 Op amp differential amplifier
(© Cengage Learning 2012)

EXAMPLE 8

Situation

Determine the input voltage difference, static voltage gain A_V, and V_{OUT} for the differential amplifier circuit of Figure 8-16.

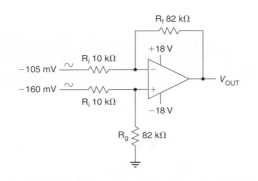

FIGURE 8-16   Op amp differential amplifier (© Cengage Learning 2012)

Solution

For Figure 8-16,

$$V_{DIFF} = V_{NONINV} - V_{INV}$$

$$= -160 \text{ mV} - (-105 \text{ mV})$$

$$= -55 \text{ mV}_{P-P}$$

$$\text{Then, } A_V = \frac{R_f}{R_i} = \frac{82 \text{ k}\Omega}{10 \text{ k}\Omega} = 8.2$$

$$\text{Then, } V_{OUT} = (V_{DIFF})(A_V)$$

$$= (-55 \text{ mV}_{P-P})(8.2)$$

$$= -451 \text{ mV}_{P-P}$$

For the circuit of Figure 8-16, the AC output voltage should be @ –451 mV$_{P-P}$ with little or no distortion because it's within the rail voltage limits of +18 V and –18 V.

Op amp differential amplifiers are used in closed-loop systems involving motor control, microphones, and twisted pair-to-coaxial cable converters for video applications.

8-7 Op Amp Summing Amplifier

The op amp **summing amplifier** provides an amplified output voltage that is the sum of two or more input voltages. The summing amplifier isn't really an amplifier in the normal sense because it doesn't multiply the input voltage(s) by a specified gain. The summing amplifier produces a larger output voltage by *adding* the input voltages.

The op amp circuits we've discussed so far have amplified AC input signals, but we mentioned in the beginning of the chapter that an op amp can also amplify DC input voltages. Thus, we will use DC input voltages in our discussion of the

summing amplifier. Figure 8-17 shows an op amp configured as a summing amplifier.

Like the non-inverting, inverting, and differential op amps discussed so far, the summing amplifier has a feedback resistor R_f connected between its output terminal and inverting terminal. The summing amplifier also has *multiple* input resistors tied to the inverting input terminal. A voltage is applied to

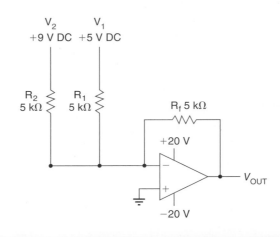

FIGURE 8-17   Op amp summing amplifier (© Cengage Learning 2012)

each input resistor, and the total, or sum, output voltage will be inverted, or flipped over.

Notice in the circuit of Figure 8-17 that there are two input resistors. A summing amplifier can have two or more input resistors. Recall that a differential amplifier has two input resistors, but these are connected to different terminals. The summing amplifier has all its resistors tied to the *same* input, the inverting terminal; thus, the output voltage will be inverted. Equation 8-8 is used to determine the output voltage of a summing amplifier.

EQUATION 8-8

Op amp summing amplifier output voltage

$$V_{OUT} = -R_f\left(\frac{V_1}{R_1} + \frac{V_2}{R_2} + \frac{V_x}{R_x} + \dots\right)$$

V_x = a third input voltage value, fourth input voltage value, etc.

R_x = a third input resistance value, fourth input resistance value, etc.

EXAMPLE 9

Situation

Determine V_{OUT} for the circuit of Figure 8-17. The input voltages are DC values.

Solution

For Figure 8-17,

$$V_{OUT} = -R_f\left(\frac{V_1}{R_1} + \frac{V_2}{R_2}\right)$$

$$V_{OUT} = -5\,k\Omega\left(\frac{5\,V}{5\,k\Omega} + \frac{9\,V}{5\,k\Omega}\right)$$

$$V_{OUT} = -5\,k\Omega\,(1\,mA + 1.8\,mA)$$

$$V_{OUT} = -5\,k\Omega\,(2.8\,mA)$$

$$V_{OUT} = -14\,V$$

Notice that this formula is just another version of Ohm's Law, V = RI. If you have only two

EXAMPLE 10

Situation

Determine V_{OUT} for the circuit of Figure 8-18.

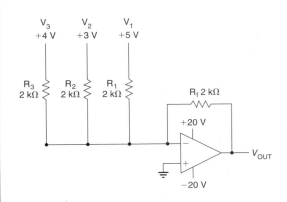

Solution

For Figure 8-18,

$$V_{OUT} = -R_f\left(\frac{V_1}{R_1} + \frac{V_2}{R_2} + \frac{V_3}{R_3}\right)$$

$$V_{OUT} = -2\,k\Omega\left(\frac{5\,V}{2\,k\Omega} + \frac{3\,V}{2\,k\Omega} + \frac{4\,V}{2\,k\Omega}\right)$$

$$V_{OUT} = -2\,k\Omega\,(2.5\,mA + 1.5\,mA + 2\,mA)$$

$$V_{OUT} = -2\,k\Omega\,(6\,mA)$$

$$V_{OUT} = -12\,V$$

FIGURE 8-18 Op amp summing amplifier (© Cengage Learning 2012)

input voltages and two input resistors, then add these values to determine the total current. If you have three input voltages and three input resistors, add all three to determine the total current. With four input voltages and input resistors, use all four. Examples 8-9 and 8-10 should clarify the process.

For the circuit of Figure 8-17, the DC output voltage should be @ −14 V with little or no distortion because this value is within the rail voltage limits of +20 V and −20 V. Note how the branch currents add up to a total current in the summing amplifier. This result is then multiplied by the R_f. Let's do another, only this time we'll use three input voltages and three input resistors.

For the circuit of Figure 8-18, the DC output voltage should be @ −12 V with little or no distortion because this value is within the rail voltage limits of +20 V and −20 V.

The summing amplifier can be used as a digital-to-analog converter. Although you may not have had a digital class yet, digital voltages are DC voltages. Figure 8-19 shows a summing amplifier used as a digital-to-analog converter.

Notice that the values of the input resistors and the feedback resistor R_f have changed. Input branches one, two, and three have *weighted* resistor values: going from right to left, with resistor R_2 being one-half the value of resistor R_1 and resistor R_3 being one-half the value of resistor R_2. This progression creates a situation where going from right to left, each branch current will double in value, which is exactly what happens in digital, or binary, systems. In digital, each position or place to the left has double the weight or value, of the position to its right. For example, the current in branch two will be double the current in branch one because the applied voltage to both branches is the same but the resistance of branch two is one-half the resistance of branch one.

What also distinguishes Figure 8-19 from previous op amp circuits is the fact that all input voltages are the same value. Finally, each branch of Figure 8-19 now has a single-pole, single throw switch (SPST) to activate or deactivate the input voltage.

How does the circuit of Figure 8-19 work? First, let's close only the switch in branch one. With SW1 closed, +5 V is applied to R_1, causing current to flow through branch one and into the op amp. The output voltage of the op amp can be determined by the following equation, a shortened version of Equation 8-8.

$$V_{OUT} = -R_f\left(\frac{V_1}{R_1}\right)$$

$$V_{OUT} = -8\,k\Omega\left(\frac{5\,V}{40\,k\Omega}\right)$$

$$V_{OUT} = -8\,k\Omega\,(.125\,mA)$$

$$V_{OUT} = -1\,V$$

With only SW1 closed, the op amp has an output voltage of −1 V.

Now, let's close both SW1 and SW2. When both switches are closed, +5 V is applied to both R_1 and R_2. Current flows through branch one and branch two and into the op amp. The output voltage of the op amp can be determined by Equation 8-8.

$$V_{OUT} = -R_f\left(\frac{V_1}{R_1} + \frac{V_2}{R_2}\right)$$

$$V_{OUT} = -8\,k\Omega\left(\frac{5\,V}{40\,k\Omega} + \frac{5\,V}{20\,k\Omega}\right)$$

$$V_{OUT} = -8\,k\Omega\,(0.125\,mA + 0.25\,mA)$$

$$V_{OUT} = -8\,k\Omega\,(0.375\,mA)$$

$$V_{OUT} = -3\,V$$

Thus, with SW1 and SW2 closed, the op amp has an output voltage of −3 V.

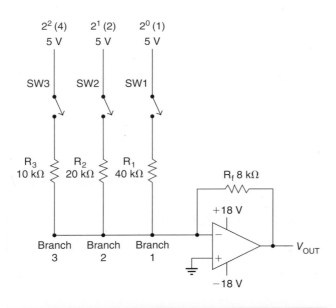

FIGURE 8-19   Op amp summing amplifier used as a digital-to-analog converter (© Cengage Learning 2012)

Now, let's close all three switches. When SW1, SW2, and SW3 are closed, $+5$ V is applied to resistors R_1, R_2, and R3. Current flows through all three branches and into the op amp. The output voltage of the op amp can be determined by an expanded version of Equation 8-8.

$$V_{OUT} = -R_f \left(\frac{V_1}{R_1} + \frac{V_2}{R_2} + \frac{V_3}{R_3} \right)$$

$$V_{OUT} = -8\,k\Omega \left(\frac{5\,V}{40\,k\Omega} + \frac{5\,V}{20\,k\Omega} + \frac{5\,V}{10\,k\Omega} \right)$$

$$V_{OUT} = -8\,k\Omega\,(.125\text{ mA} + 0.25\text{ mA} + .5\text{ mA})$$

$$V_{OUT} = -8k\Omega\,(.875\text{ mA})$$

$$V_{OUT} = -7\text{ V}$$

Thus, if all three switches are closed in the circuit of Figure 8-19, the op amp will have an output voltage of -7 V. The reason the V_{OUT} is negative in all three cases is because the current from any or all of the branches enters the op amp's inverting terminal, which changes the polarity. Now it's time to build an op amp summing amplifier on the breadboard.

LAB ACTIVITY 8-3

Op Amp Summing Amplifier

Materials, Equipment, and Parts:

- *NTE* catalog or Internet access, www.nteinc.com

- PC w/Multisim®, Electronics Workbench®, or SPICE.

- DMM with test leads.

- One positive DC voltage source, 5V.

- One positive DC voltage source, 0–20 V.

- One negative DC voltage source, 0–20 V.

- Breadboard and connecting wires.

- LM741CN, UA741CN, or equivalent op amp IC.

- Fixed resistors: 4 kΩ, 5 kΩ, 10 kΩ, and 20 kΩ.

- SPST switch (3) or SPDT switch (3).

Discussion Summary:

The op amp summing amplifier used as a digital-to-analog converter provides a DC output voltage that corresponds to one or more DC input voltages. If input resistors are arranged in weighted positions, the input circuit behaves like a digital system, and the op amp produces an analog output voltage.

Procedure:

SAFETY FIRST. Eye protection should always be worn when working with live voltages. Before powering on a live circuit, always check with your instructor.

1 Write the part number of the op amp at your workstation. Then use the *NTE* catalog or visit the website www.nteinc.com to find and record the NTE replacement number, diagram number, and V_{CC}.

Part number _____

NTE number _____

Diagram # _____

V_{CC} _____

2 Draw the op amp pin configuration diagram and label each pin.

(continues)

LAB ACTIVITY 8-3

(continued)

3 Build the circuit of Figure 8-20 on your breadboard.

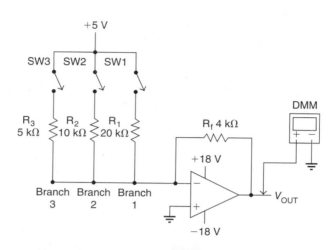

FIGURE 8-20   Op amp summing amplifier used as a digital-to-analog converter
(© Cengage Learning 2012)

4 Connect the positive lead of the +18 V DC voltage source to pin 7 of the op amp.

5 Connect the positive lead of the −18 V DC voltage source to pin 4 of the op amp.

6 Connect one terminal of each SPST switch to the +5 V DC voltage source.

7 Connect the other terminal of each SPST switch to one end of each input resistor.

8 Connect the other end of all three input resistors to the inverting terminal of the op amp, pin 2.

9 Set the DMM to measure DC volts. Connect the positive lead of the DMM to the output terminal of the op amp, pin 6.

10 Connect the black lead of the +18 V DC voltage source, the black lead of the −18 V DC voltage source, the black lead of the +5 V DC voltage source, the black lead of the DMM, and a wire from the non-inverting terminal, pin 3, to one point (hole) on the breadboard, which is ground.

11 Have the instructor check your circuit.

12 Calculate and record the V_{OUT} of the op amp when *only* SW1 in branch one is closed.

$$V_{OUT} = -R_f \left(\frac{V_1}{R_1} \right)$$

V_{OUT} _____

LAB ACTIVITY 8-3

13 Power on the +18 V DC voltage source, the −18 V DC voltage source, the +5 V DC voltage source, and the DMM. Turn on SW1. Record the measured V_{OUT} value below.

V_{OUT} _____

Power off.

How does the measured V_{OUT} compare to the calculated V_{OUT} of Step 12?

14 Calculate and record the V_{OUT} of the op amp when both SW1 and SW2 are closed.

$$V_{OUT} = -R_f \left(\frac{V_1}{R_1} + \frac{V_2}{R_2} \right)$$

V_{OUT} _____

15 Power on the +18 V DC voltage source, the −18 V DC voltage source, the +5 V DC voltage source, and the DMM. Turn on SW1 and SW2. Record the measured V_{OUT} value below.

V_{OUT} _____

Power off.

How does the measured V_{OUT} compare to the calculated V_{OUT} of Step 14?

16 Calculate and record the V_{OUT} of the op amp when SW1, SW2, and SW3 are closed.

$$V_{OUT} = -R_f \left(\frac{V_1}{R_1} + \frac{V_2}{R_2} + \frac{V_3}{R_3} \right)$$

V_{OUT} _____

17 Power on the +18 V DC voltage source, the −18 V DC voltage source, the +5 V DC voltage source, and the DMM. Turn on SW1, SW2, and SW3. Record the measured V_{OUT} value below.

V_{OUT} _____

Power off.

How does the measured V_{OUT} compare to the calculated V_{OUT} of Step 16?

18 Power on the +18 V DC voltage source, the −18 V DC voltage source, the +5 V DC voltage source, and the DMM. Turn on only SW3 and SW1. What is the measured V_{OUT}? Why is it this value? Power off.

19 Have the instructor verify your results.

20 Build the circuit on Multisim®, Electronics Workbench®, or SPICE and then compare the results.

8-8 Fine-Tuning and Troubleshooting Op Amps

Now that we've analyzed the six basic op amp configurations—comparator, voltage follower, non-inverting amplifier, inverting amplifier, differential amplifier, and summing amplifier—it's time to look at two minor parts of the op amp IC: the offset null pins and the internal capacitor.

Recall that the offset null pins are used only for applications involving sensitive measurements, such as the instruments used in offshore oil and gas drilling. When the positive and negative DC supply voltage terminals provide power to an op amp IC, the internal transistors and resistors drop voltages. These internal transistors and resistors are not perfectly aligned, so the output pin of an op amp may show a reading in the μV or mV range *with no input voltage(s) applied*. Figure 8-21(a) shows this condition. For sensitive measurements, this output voltage could show a false reading when input voltages are then applied.

To eliminate this problem, a tiny variable resistor (@ 10 kΩ) is soldered between the offset null pins. The middle terminal of the variable resistor is then soldered to the negative DC supply voltage pin.

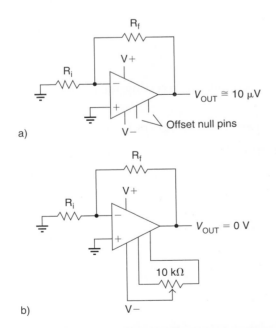

a)

b)

FIGURE 8-21 Op amp output voltage with no input voltages applied: a) No offset null adjustment and b) Offset null adjustment (© Cengage Learning 2012)

With DC power applied to the op amp, this resistor is adjusted until the output of the op amp reads zero, or null. Then, glue is applied to the variable resistor to prevent it from turning. The variable resistor is shown in Figure 8-21(b). This process keeps the op amp output at 0 V when no input voltage(s) is applied. Thus, the offset null pins are used to adjust or calibrate the output of an op amp to 0 volts when it has no input voltage(s) applied. I actually performed this task on MWD (Measurement While Drilling) instruments when I worked for an oilfield service company. Again, this is a rare application involving the offset null pins. Most op amp applications don't use the pins at all.

Looking way back at Figure 8-1, we see that the 741 op amp IC contains an internal capacitor. This capacitor is called a **frequency compensation capacitor**, and it is used to limit the high-frequency operation of the op amp. As frequency increases, the capacitive reactance of the capacitor decreases to a point where eventually the gain of the op amp is 1, which is really no gain at all.

The procedure for troubleshooting an op amp is basically the same as for any IC:

1) Perform a visual inspection.
2) Check that the DC power supply voltages (V+ and V−) and ground are correct.
3) Check R_f and R_i.
4) Check the input voltage(s).
5) Check the output voltage.

If Steps 1 through 4 check okay and Step 5 shows no output voltage, then the IC is bad and must be replaced.

So far, we've analyzed all six basic op amp configurations and built three of them on the breadboard. Besides the applications we've already discussed, op amps are used in hundreds of applications, including square wave and sine wave generators, low-pass and high-pass filters in radio receivers, CD players, transducers, strain gages, calculators, ADCs, instrumentation amplifiers, and many more.

Chapter Summary

The op amp is a high-gain differential amplifier that can amplify AC or DC voltages. It has a high input impedance of @ 300 kΩ to 2 MΩ and an output impedance of @ 100 Ω. The open-loop gain of an op amp (no feedback circuitry) can vary from 100,000 to 1 million. Early op amps were used in analog computers for mathematical operations such as addition and subtraction, but today they are used in hundreds of applications, including square wave and sine wave generators, low-pass and high-pass filters in radio receivers, CD players, transducers, strain gages, calculators, ADCs, and instrumentation amplifiers.

The standard op amp is the 741 IC, which has 35 internal devices including transistors, resistors, one diode, and one capacitor. A technician has access to only the eight external pins: the inverting terminal, used to amplify and invert an AC input voltage or change the polarity of a DC input voltage; the non-inverting input terminal, which provides only amplification; the positive and negative DC supply voltage terminals, which provide operating power to the op amp; two offset null pins, used for calibrating the op amp in applications involving sensitive measurements; and one pin for the output. The eighth pin has no connection.

A technician connects resistors to the external pins to control (reduce) the gain of an op amp. Negative feedback through resistor R_f reduces the gain. There are six basic op amp configurations: comparator, voltage follower, non-inverting amplifier, inverting amplifier, differential amplifier, and summing amplifier. The comparator configuration uses no external feedback circuitry. The voltage follower, non-inverting amplifier, inverting amplifier, differential amplifier, and summing amplifier all use some type of external feedback circuitry.

For any op amp circuit, the *maximum* AC output signal voltage (V_{RMS}) will typically be one or two volts lower than the rail voltage. Like all amplifiers, driving an op amp beyond saturation can result in distortion or clipping of the input waveform. This makes the open-loop op amp configuration impractical without additional feedback circuitry to control the gain. For this reason, op amps usually use additional circuitry involving feedback to control the gain.

The op amp voltage follower has a gain of 1 and is used only to match a high impedance stage to a low impedance stage, such as an audio amplifier to one or more loudspeakers.

The op amp non-inverting amplifier uses resistors that provide negative feedback to control the voltage gain. The non-inverting op amp provides an amplified voltage and is used in video amplifiers for cable televisions and low-voltage display screens used on cell phones.

The op amp inverting amplifier also uses external resistors to provide negative feedback and control the op amp voltage gain. An op amp inverting amplifier both inverts and amplifies an AC signal. Applications for an inverting op amp include photodiode and phototransistor amplifiers in solar cells and panels.

The op amp differential amplifier provides an amplified output voltage that is the difference of two input voltages. The differential amplifier works like the comparator except that it has external resistors to provide negative feedback and control the voltage gain. Applications for the op amp differential amplifier include closed-loop systems that involve motor control.

The op amp summing amplifier provides an amplified output voltage that is the sum of two or more input voltages. The summing amplifier is used in applications such as DACs and CD players.

The procedure for troubleshooting op amps includes visual inspection, checking power and ground pins, checking resistors, checking input voltage(s), and then checking the output voltage.

CHAPTER EQUATIONS

Comparator open-loop input voltage

(Equation 8-1)

$$V_{IN} = V_{NONINV} - V_{INV}$$

V_{NONINV} = voltage at non-inverting terminal

V_{INV} = voltage at inverting terminal

Non-inverting op amp static voltage gain

(Equation 8-2)

$$A_V = \frac{R_f}{R_i} + 1$$

R_f = feedback resistance

R_i = input resistance

Inverting op amp static voltage gain

(Equation 8-3)

$$A_V = \frac{R_f}{R_i}$$

R_f = feedback resistance

R_i = input resistance

Inverting op amp grounding resistor

(Equation 8-4)

$$R_g = \frac{(R_f)(R_i)}{R_f + R_i}$$

R_f = feedback resistance

R_i = input resistance

Differential op amp input voltage difference

(Equation 8-5)

$$V_{DIFF} = V_{NONINV} - V_{INV}$$

V_{NONINV} = voltage at non-inverting terminal

V_{INV} = voltage at inverting terminal

Differential op amp static voltage gain

(Equation 8-6)

$$A_V = \frac{R_f}{R_i}$$

R_f = feedback resistance

R_i = either input resistance

Differential op amp output voltage

(Equation 8-7)

$$V_{OUT} = (V_{DIFF})(A_V)$$

Op amp summing amplifier output voltage

(Equation 8-8)

$$V_{OUT} = -R_f \left(\frac{V_1}{R_1} + \frac{V_2}{R_2} + \frac{V_X}{R_X} + \cdots \right)$$

V_x = a third input voltage value, fourth input voltage value, etc.

R_x = a third input resistance value, fourth input resistance value, etc.

CHAPTER REVIEW QUESTIONS

Chapter 8-1

1. Define operational amplifier.

2. What makes an op amp unusual in terms of amplification capability?

3. What are the typical input and output impedances of an op amp?

4. In what applications were the first op amps used?

5. What generic part number is used for an op amp IC?

6. Draw the schematic symbol for a 5-terminal op amp and explain the function of each terminal.

7. List the six basic op amp configurations.

Chapter 8-2

8. What does an op amp's open-loop voltage gain mean?

9. What are an op amp's rail voltages?

10. If an op amp has a rail voltage of ±18 V, what is its maximum output voltage?

11. Determine V_{IN} and V_{OUT} for the circuit of Figure 8-22. Use 100,000 for the A_{OL}.

```
          +18 V
70 μV_P-P  ~ |‾\
             |  \
100 μV_P-P ~ |  /— V_OUT
             |_/   A_OL ≈ 100,000
          −18 V
```

FIGURE 8-22 Op amp comparator configured for open-loop voltage gain (© Cengage Learning 2012)

12. Why is an op amp configured for open-loop gain impractical?

Chapter 8-3

13. What is the A_V of the op amp voltage follower?

14. Name one application for an op amp voltage follower.

Chapter 8-4

15. How does one control the gain of an op amp?

16. Does negative feedback increase or decrease the gain of an op amp?

17. What do you call the gain of an op amp that uses feedback circuitry?

18. Determine the static voltage gain and output voltage for the op amp non-inverting amplifier in Figure 8-23.

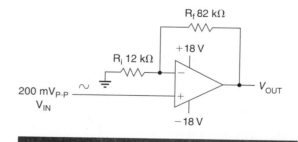

FIGURE 8-23 Op amp non-inverting amplifier (© Cengage Learning 2012)

Chapter 8-5

19. The op amp inverting amplifier produces a 180° phase shift of the AC input signal. Which BJT amplifier configuration also produces a 180° phase shift of the AC input signal?

20. Determine the static voltage gain and output voltage for the op amp inverting amplifier in Figure 8-24.

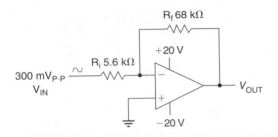

FIGURE 8-24 Op amp inverting amplifier (© Cengage Learning 2012)

Chapter 8-6

21. Determine the input voltage difference, static voltage gain, and output voltage for the differential amplifier circuit of Figure 8-25.

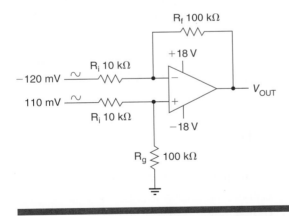

FIGURE 8-25 Op amp differential amplifier (© Cengage Learning 2012)

Chapter 8-7

22. Determine the V_{OUT} of the summing amplifier used for digital-to-analog conversion in Figure 8-26 if only switches SW3 and SW2 are closed.

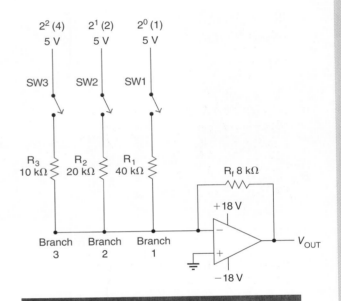

FIGURE 8-26 Op amp summing amplifier used as a digital-to-analog converter (© Cengage Learning 2012)

23. Explain the purpose of the single capacitor in an op amp IC.

24. What are an op amp's offset null pins used for?

25. Name five common applications for op amps.

Analog Oscillators

OBJECTIVES   *Upon completion of this chapter, you should be able to:*

- Define oscillator and list the four requirements for oscillation.
- Identify a Hartley oscillator and explain its operation.
- Distinguish between positive and negative feedback.
- Calculate the frequency of oscillation for the tank section of a Hartley oscillator.
- Build a Hartley oscillator to calculate and measure its frequency of oscillation.
- Identify a Colpitts oscillator and explain its operation.
- Calculate the frequency of oscillation for the tank section of a Colpitts oscillator.

- Build a Colpitts oscillator to calculate and measure its frequency of oscillation.
- Draw and label the schematic symbol for a crystal oscillator.
- Explain the concept of piezoelectricity.
- List and explain the factors that determine the resonant frequency of a crystal.
- Explain the differences between LC oscillators (Hartley and Colpitts) and crystal oscillators in terms of frequency accuracy and stability.
- Discuss the two methods used to create additional frequencies from one crystal oscillator.

MATERIALS, EQUIPMENT, AND PARTS

Materials, equipment, and parts needed for the lab experiments in this chapter are listed below:

- *NTE* catalog or Internet access, www.nteinc.com (*Look it up to hook it up.*)
- PC w/Multisim®, Electronics Workbench®, or SPICE.
- DMM with test leads.
- Single trace oscilloscope w/BNC-to-alligator leads.
- 5 V DC voltage source.

- 12 V DC voltage source.
- Breadboard and connecting wires.
- 2N3904 or equivalent NPN transistor.
- 0.01 µF fixed capacitor, 0.1 µF fixed capacitor (3), and 0.22 µF fixed capacitor.

- 4.7 mH inductor coil (2), 10 mH inductor coil, and 100 mH inductor coil.
- 150 Ω fixed resistor, 1 kΩ fixed resistor, 10 kΩ fixed resistor (2), and 33 kΩ fixed resistor.

GLOSSARY OF TERMS

Analog oscillator An electronic circuit that creates an AC sine wave from a DC voltage source and a tank circuit (or crystal)

Frequency of oscillation, or f_o The frequency created by the tank circuit of an analog oscillator

Hartley oscillator An LC oscillator that has two coils (or a center-tapped coil) and one capacitor in its tank section

Flywheel effect The charging/discharging action between the coil(s) and capacitor(s) of the tank circuit of an LC oscillator that creates a sine wave

Positive feedback A form of feedback where the feedback signal is in phase with the input signal, which increases the gain of an amplifier circuit

Colpitts oscillator An LC oscillator that has two capacitors and one coil in its tank section

Clapp oscillator An LC oscillator that has three capacitors and one coil in its tank section; the third capacitor reduces the effects of rogue capacitance, giving the oscillator better frequency stability than the Hartley or Colpitts oscillators

Crystal oscillator A high-precision oscillator that typically uses a quartz crystal to create a sine wave

Piezoelectricity The property of certain rock crystals such as quartz and Rochelle salt that creates a sine wave when an electromotive force (voltage) is applied to it

9-1 GENERAL PRINCIPLES OF OSCILLATION

An **analog oscillator** is an electronic circuit that creates an AC sine wave from a DC voltage source and a tank circuit (or crystal). Four conditions are required for oscillation: a DC voltage source; a tank circuit or crystal; positive, or regenerative, feedback; and amplification.

You may recall from Lab Activity 6-2 that the class C amplifier had poor fidelity because it was hardly on (<180° of the AC input signal) and that the amplifier reproduced only a small part of the AC input signal. Remember, too, that we connected a tank section to the transistor's collector to restore the distorted AC waveform. Review the "Link to Prior Learning" below about the tank section.

Link to Prior Learning

A tank section contains a coil and a capacitor in parallel. The capacitor can be a fixed type or a variable type. If the capacitor is variable, it can be adjusted to tune the section to a desired frequency, called the resonant frequency (f_r). A tank section with an adjustable capacitor is called a tuned circuit. Tuned circuits are used at the input of radio receivers or at the output of radio transmitters.

Well, the tank section is also at the heart of an analog oscillator; however, instead of restoring an AC waveform, the oscillator *creates* one. We can build oscillator circuits by choosing values of inductance and capacitance that will produce a desired output frequency, which we'll now call the **frequency of oscillation (f_o)**. The tank section of the Class C amplifier contained only one capacitor and one coil, or inductor. In the following sections, we'll see that the tank section can be made of various combinations of coils and capacitors. We'll also revisit a formula from Chapter 6 and apply it to oscillators.

9-2 HARTLEY OSCILLATOR

The **Hartley oscillator** is a sine-wave oscillator that has two coils (or a center-tapped coil) and one capacitor in its tank section. Figure 9-1 shows a Hartley oscillator.

The two major parts of the Hartley oscillator are the tank section and the amplifier section. Each section produces a 180° phase shift of the AC signal voltage, so by the time the waveform leaves the oscillator, it looks like a standard sine wave: the positive alternation comes first, and the negative alternation follows.

The tank section of Figure 9-1 has two coils, L_1 and L_2, and one capacitor, C_1. These are the key visual features of the Hartley oscillator. Although a wire is connected between the two coils, one center-tapped coil can also be used. *This tank section, or circuit, of an LC oscillator determines the output frequency.*

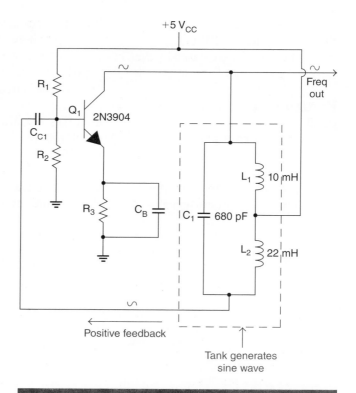

FIGURE 9-1   Hartley oscillator (© Cengage Learning 2012)

How exactly does a Hartley oscillator work? In Figure 9-1, the 5 V DC voltage source connected between L_1 and L_2 jump starts the tank circuit. Capacitor C_1 in the tank charges and then discharges through coils L_1 and L_2, creating a back-and-forth or up-and-down voltage waveform—a sine wave. This action between the coils and capacitor is also called the **flywheel effect**, or washing machine effect, because of the back-and-forth motion. The sine wave is small and leads with its negative alternation; it will eventually die out unless it's amplified.

This is when transistor Q_1 comes into play. The sine wave created by the tank circuit is coupled by C_{C1} to the base of transistor Q_1. Notice that Q_1 is configured as a common-emitter, so it takes the input signal created by the tank circuit and inverts it to a standard sine wave with a leading positive alternation. From Q_1, the signal travels to whatever load is to be used. Thus, Q_1 provides the amplification and inversion to increase and correct the sine wave that was created by the tank circuit.

INTERNET ALERT

Check the website http://www.falstad.com/circuit/e-hartley.html to see the operation of a Hartley oscillator.

The Hartley oscillator of Figure 9-1 satisfies all conditions for oscillation; it has a DC voltage source, a tank circuit, positive feedback, and amplification. Put simply, **positive feedback** in a system means that the feedback signal is in phase with the input signal (both are reaching their positive and negative peaks at the same time). Recall from Chapter 8 that the op amp uses *negative* feedback to control the gain by reducing it. Positive feedback, however, increases the gain of a system. Because the tank circuit creates the input signal that is fed to the amplifier, *positive* feedback in an oscillator circuit really means feedback that is *not* negative.

Ralph Vinton Lyon Hartley invented the vacuum tube version of the Hartley oscillator at Western Electric Company in 1915. The solid-state version is still used in many applications today: signal generators that create frequencies from 1 Hz to 500 MHz; radio frequency applications such as broadcast radio transmitters and receivers operating in the 540 kHz to 1630 kHz range (AM) and 88 MHz to 108 MHz range (FM); and many other electronic communication systems. The advantages of the Hartley oscillator include its simplicity of design and its ability to create a fairly constant output voltage over a large range of frequencies. The major disadvantages of the Hartley oscillator involve frequency accuracy and frequency stability, which we'll discuss shortly.

Since the tank section determines the operating frequency of an oscillator, we'll review the equation that is used to calculate the resonant frequency of a tank section.

EQUATION 6-2

Class C amplifier resonant frequency of tank section

$$f_r = \frac{1}{2\pi\sqrt{LC}}$$

f_r = resonant frequency in Hz

L = inductance of coil in tank

C = capacitance of capacitor in tank

This same equation can be used to calculate the frequency of oscillation for a Hartley oscillator. First, we have to modify the equation slightly because we have two coils in the tank circuit.

Link to Prior Learning

Inductors (coils) in series

Inductors in series are like resistors in series, so the total inductance equals the sum of the individual inductances.

$$L_T = L_1 + L_2$$

The equation for Hartley oscillators is Equation 9-1.

EQUATION 9-1

Hartley oscillator frequency of oscillation

$$f_o = \frac{1}{2\pi\sqrt{L_T C}}$$

f_o = frequency of oscillation in Hz

L_T = total inductance of coils in tank
    (if more than one coil)

C = capacitance of capacitor in tank

EXAMPLE 1

Situation

What is the frequency of oscillation for the Hartley oscillator in Figure 9-1?

Solution

$$L_T = L_1 + L_2 = 10 \text{ mH} + 22 \text{ mH}$$

$$= 32 \text{ mH}$$

$$f_o = \frac{1}{2\pi\sqrt{L_T C}}$$

$$= \frac{1}{2\pi\sqrt{(32 \text{ mH})(680 \text{ pF})}}$$

$$= 34{,}135.86 \text{ Hz}$$

When using the frequency of oscillation equation, work from right to left and bottom to top. For example, first find the total inductance of the tank circuit, multiply this result by the capacitor value, take the square root of this answer, multiply the new answer by 2π (6.28), and then use the invert function on your calculator.

Thus, the Hartley oscillator in Figure 9-1 will generate a frequency of @ 34.13 kHz. Let's do another.

The Hartley oscillator in Figure 9-2 will generate a frequency of about 1.12 kHz. We say "about" because the Hartley oscillator doesn't generate an exact frequency. The calculated frequency won't always match the measured frequency because the coils and capacitors that make up the tank circuit aren't exact values. For example, a 1 μF capacitor might actually measure anywhere between 0.9 μF and 1.1 μF, and a small change in capacitance can cause a large change in the frequency of oscillation.

Frequency stability is also an issue with the Hartley oscillator. Over time, the frequency drifts from its calculated or nominal value due to the deterioration of the capacitors and coils. Because

of these factors, oscillators—like many electronic devices—have frequency tolerances. The Hartley oscillator has a frequency tolerance typically between ±0.01% and ±0.1%. This means that for a calculated f_o of 10,000 Hz, the measured frequency of the Hartley oscillator could be between 9,990 Hz and 10,010 Hz. Don't be surprised if the difference is more. Keep this in mind when you're building a Hartley oscillator circuit: the calculated frequency of oscillation and measured frequency of oscillation don't have to match exactly.

Now that we've analyzed the operation of a Hartley oscillator and calculated its tank frequency of oscillation, let's build it on the breadboard.

EXAMPLE 2

Situation

What is the frequency of oscillation for the Hartley oscillator in Figure 9-2?

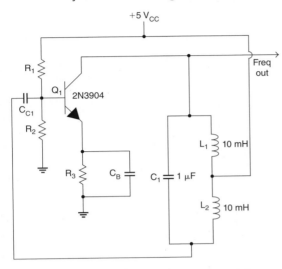

FIGURE 9-2   Hartley oscillator (© Cengage Learning 2012)

Solution

$$L_T = L_1 + L_2 = 10 \text{ mH} + 10 \text{ mH}$$
$$= 20 \text{ mH}$$

$$f_o = \frac{1}{2\pi\sqrt{L_T C}} = \frac{1}{2\pi\sqrt{(20 \text{ mH})(1 \text{ } \mu\text{F})}}$$
$$= 1125.96 \text{ Hz}$$

LAB ACTIVITY 9-1

Hartley Oscillator

Materials, Equipment, and Parts:

- *NTE* catalog or Internet access, www.nteinc.com (*Look it up to hook it up.*)

- PC w/Multisim®, Electronic Workbench®, or SPICE.

- DMM with test leads.

- Single trace oscilloscope w/BNC-to-alligator leads.

- 5 V DC voltage source.

- Breadboard and connecting wires.

- 2N3904 or equivalent NPN transistor.

- 0.1 μF fixed capacitor (3).

- 4.7 mH inductor coil (2).

- 1 kΩ fixed resistor and 10 kΩ fixed resistor (2).

Discussion Summary:

A Hartley oscillator creates an AC sine wave from a tank circuit that is made of two coils and a capacitor or one center-tapped coil and a capacitor. The tank circuit coil and capacitor values determine the output frequency. The inverted output waveform from the tank circuit is applied to a transistor, which amplifies the waveform and inverts it to a standard sine wave with a leading positive alternation.

Procedure:

SAFETY FIRST. Eye protection should always be worn when working with live voltages. Before powering on a live circuit, always check with your instructor.

1 Write the part number of the transistor at your workstation. Then use the *NTE* catalog or visit the website www.nteinc.com to find and record the NTE replacement number and the diagram number.

Part number _____

NTE number _____

Diagram # _____

2 Draw the transistor outline and label the emitter, base, and collector for the transistor.

(continues)

LAB ACTIVITY 9-1

(continued)

3 Build the circuit of Figure 9-3 on your breadboard.

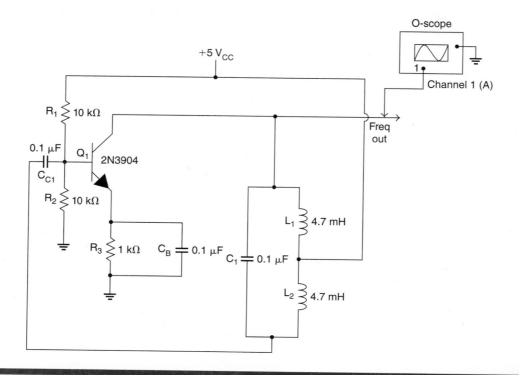

FIGURE 9-3   Hartley oscillator (© Cengage Learning 2012)

4 Calculate the frequency of oscillation for the Hartley oscillator circuit.

$$L_T = L_1 + L_2 =$$

$$f_o = \frac{1}{2\pi\sqrt{L_T C}} =$$

$$f_o \ _____$$

5 Connect the positive lead of the 5 V DC voltage source to the top end of resistor R_1 and also between inductor L_1 and L_2.

6 Connect the positive lead of the oscilloscope's Channel 1 to the collector of Q_1 or to the top end of the tank circuit.

LAB ACTIVITY 9-1

7   Connect the black lead of the 5 V DC voltage source, the black lead of the o-scope, a wire from the bottom of resistor R_2, and a wire from the bottom of resistor R_3 to one point (hole) on the breadboard, which is ground.

8   Have the instructor check your circuit.

9   Power on the 5 V DC voltage source and the o-scope.

10   The oscillator should start conducting, and a waveform should appear on your o-scope within ten seconds. You may have to adjust your oscilloscope volts/division settings and time/division settings so that the AC input waveform appears on your o-scope display. Describe the appearance of the waveform below.

11   Determine the period of the waveform that appears on your o-scope.

$$p = \frac{\text{\# horizontal divisions}}{1} \times \frac{\text{TIME}}{\text{DIV}} =$$

Now, use the period to determine the frequency of the waveform.

$$f_o = \frac{1}{p} =$$

f_o _____

12   How does the measured frequency of oscillation in Step 11 compare to the calculated frequency of oscillation in Step 4? Explain any significant differences.

13   Power off the 5 V DC source and the o-scope. Set your DMM to measure frequency. Connect the positive lead of the DMM to the collector of Q_1 or to the top end of the tank circuit. Connect the negative end of your DMM to ground. Power on the 5 V DC source and your DMM. Measure and record the frequency.

f_o _____

14   How does the measured frequency of oscillation in Step 13 compare to the calculated frequency of oscillation in Step 4.

15   Have the instructor verify your results.

16   Build the circuit on Multisim®, Electronics Workbench®, or SPICE and then compare the results.

9-3 COLPITTS OSCILLATOR

The **Colpitts oscillator** is a sine-wave oscillator that has two capacitors and one coil in its tank section. Figure 9-4 shows a Colpitts oscillator.

Like the Hartley oscillator, the two major parts of the Colpitts oscillator are the tank section and the amplifier section. Each section produces a 180° phase shift of the AC signal voltage, so by the time the waveform leaves the oscillator, it looks like a standard sine wave: the positive alternation comes first, and the negative alternation follows.

Notice the tank section of Figure 9-4 has two capacitors, C_1 and C_2, and one coil. In addition, a wire is connected between the two capacitors and ground. These are the key visual features of the Colpitts oscillator. *Again, the tank section, or circuit, of an LC oscillator determines the output frequency.*

The operation of a Colpitts oscillator is very similar to the operation of the Hartley oscillator. In Figure 9-4, the 12 V DC voltage source connected to the top of the tank circuit jump starts the tank circuit. Capacitors C_1 and C_2 in the tank charge and then discharge through coil L_1, creating a back-and-forth or up-and-down voltage waveform—a sine wave. Again, this action between the capacitors and coil is called the fly-wheel effect or washing machine effect because of the back-and-forth motion. The sine wave is small and leads with its negative alternation; it will eventually die out unless it's amplified.

This is when transistor Q_1 comes into play. The sine wave created by the tank circuit is coupled by C_{C1} to the base of transistor Q_1. As in the Hartley oscillator, Q_1 is configured as a common-emitter, and thus it takes the input signal created by the tank circuit and inverts it to a standard sine wave with a leading positive alternation before the signal goes to whatever load is being used. Thus, Q_1 provides the amplification and inversion to increase and correct the sine wave that was created by the tank circuit.

The Colpitts oscillator of Figure 9-4 satisfies all conditions for oscillation; it has a DC voltage source, a tank circuit, positive feedback, and amplification. Edwin H. Colpitts invented the vacuum tube version of the Colpitts oscillator at Western Electric Company in 1919. In fact, Colpitts and Hartley were co-workers. Like the Hartley oscillator, the solid-state version of the Colpitts oscillator is still used in many applications today: signal generators, radio frequency applications such as broadcast radio receivers and transmitters, and other electronic communication systems. The advantages of the Colpitts oscillator include the simplicity of design and the ability to create a fairly constant output voltage over a large range of frequencies. Like the Hartley, the main disadvantage of the Colpitts oscillator is that its output frequency is not precise and is subject to drift.

As with the Hartley oscillator, the resonant frequency equation can be used to calculate the frequency of oscillation for a Colpitts. First, we have to modify the equation slightly because we have two capacitors in the tank circuit.

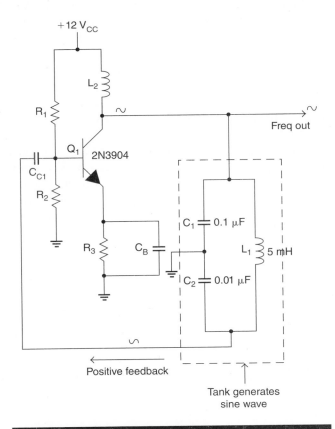

FIGURE 9-4 Colpitts oscillator (© Cengage Learning 2012)

Link to Prior Learning

Capacitors in series

Capacitors in series are like resistors in parallel, so the total capacitance equals the product of the capacitors divided by the sum of the capacitors.

$$C_T = \frac{(C_1)(C_2)}{C_1 + C_2}$$

The equation for Colpitts oscillators is Equation 9-2.

EQUATION 9-2

Colpitts oscillator frequency of oscillation

$$f_o = \frac{1}{2\pi\sqrt{LC_T}}$$

f_o = frequency of oscillation in Hz

L = inductance of coil in tank

C_T = total capacitance of capacitors in tank

When using the frequency of oscillation equation, work from right to left and bottom to top. For example, first find the total capacitance of the tank circuit, multiply this result by the coil value, take the square root of this answer, multiply the new answer by 2π (6.28), and then use the invert function on your calculator.

Thus, the Colpitts oscillator in Figure 9-4 will generate a frequency of @ 23.61 kHz. Let's do another.

EXAMPLE 3

Situation

What is the frequency of oscillation for the Colpitts oscillator in Figure 9-4?

Solution

$$C_T = \frac{(C_1)(C_2)}{C_1 + C_2} = \frac{(0.1 \ \mu F)(0.01 \ \mu F)}{0.1 \ \mu F + 0.01 \ \mu F}$$

$$= 9.09 \ nF$$

$$f_o = \frac{1}{2\pi\sqrt{LC_T}} = \frac{1}{2\pi\sqrt{(5 \ mH)(9.09 \ nF)}}$$

$$= 23,619.64 \ Hz$$

Thus, the Colpitts oscillator in Figure 9-5 will generate a frequency of @ 87.66 kHz. The frequency accuracy and stability is a bigger issue with the Colpitts than the Hartley because the Colpitts has two capacitors in its tank circuit. Also, the leads between the capacitors and coils and other parts of the feedback circuit have their own capacitance, which affect the overall frequency of oscillation. Like the Hartley, a Colpitts oscillator has a frequency tolerance typically between ±0.01% and ±0.1%. Again, keep this in mind when you're building a Colpitts oscillator circuit: the calculated frequency of oscillation and the measured frequency of oscillation don't have to match exactly.

Let's build the Colpitts oscillator on the breadboard.

EXAMPLE 4

Situation

What is the frequency of oscillation for the Colpitts oscillator in Figure 9-5?

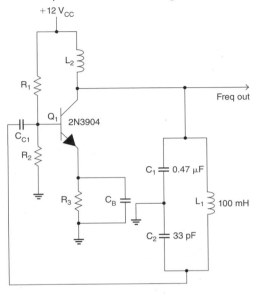

FIGURE 9-5 Colpitts oscillator (© Cengage Learning 2012)

Solution

$$C_T = \frac{(C_1)(C_2)}{C_1 + C_2} = \frac{(0.47\,\mu F)(33\,pF)}{0.47\,\mu F + 33\,pF}$$

$$= 32.99\,pF$$

$$f_o = \frac{1}{2\pi\sqrt{LC_T}}$$

$$= \frac{1}{2\pi\sqrt{(100\,\text{mH pF})(32.99\,pF)}}$$

$$= 87{,}669.63\,\text{Hz}$$

LAB ACTIVITY 9-2

Colpitts Oscillator

Materials, Equipment, and Parts:

- *NTE* catalog or Internet access, www.nteinc. com (*Look it up to hook it up.*)

- PC w/Multisim®, Electronics Workbench®, or SPICE.

- DMM with test leads.

- Single trace oscilloscope w/BNC-to-alligator leads.

- 12 V DC voltage source.

- Breadboard and connecting wires.

- 2N3904 or equivalent NPN transistor.

- 0.01 μF fixed capacitor, 0.1 μF fixed capacitor (2), and 0.22 μF fixed capacitor.

- 10 mH inductor coil and 100 mH inductor coil.

- 150 Ω fixed resistor, 10 kΩ fixed resistor, and 33 kΩ fixed resistor.

Discussion Summary:

A Colpitts oscillator creates an AC sine wave from a tank circuit that is made of two capacitors and one coil. The tank circuit coil and capacitor values determine the output frequency. The inverted output waveform from the tank circuit is applied to a transistor, which amplifies the waveform and inverts it to a standard sine wave with a leading positive alternation.

Procedure:

SAFETY FIRST. Eye protection should always be worn when working with live voltages. Before powering on a live circuit, always check with your instructor.

1 Write the part number of the transistor at your workstation. Then use the *NTE* catalog or visit the website www.nteinc.com to find and record the NTE replacement number and the diagram number.

Part number _____

NTE number _____

Diagram # _____

2 Draw the transistor outline and label the emitter, base, and collector for the transistor.

(continues)

LAB ACTIVITY 9-2

(continued)

3   Build the circuit of Figure 9-6 on your breadboard.

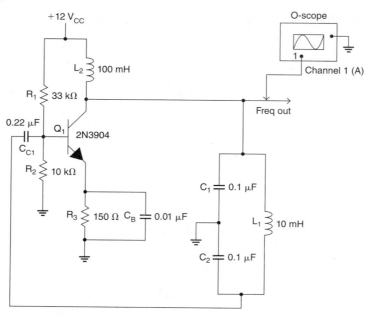

FIGURE 9-6   Colpitts oscillator (© Cengage Learning 2012)

4   Calculate the frequency of oscillation for the Colpitts oscillator circuit.

$$C_T = \frac{(C_1)(C_2)}{C_1 + C_2}$$

$$f_o = \frac{1}{2\pi\sqrt{LC_T}} =$$

f_o _____

5   Connect the positive lead of the 12 V DC voltage source to the top end of resistor R_1 and to the top of coil L_2.

6   Connect the positive lead of the oscilloscope's Channel 1 to the collector of Q_1 or to the top end of the tank circuit.

7   Connect the black lead of the 12 V DC voltage source, the black lead of the o-scope, a wire from the bottom of resistor R_2, a wire from the bottom of resistor R_3, and a wire from the center of C_1 and C_2 in the tank to one point (hole) on the breadboard, which is ground.

LAB ACTIVITY 9-2

8 Have the instructor check your circuit.

9 Power on the 12 V DC voltage source and the o-scope.

10 The oscillator should start conducting, and a waveform should appear on your o-scope within ten seconds. You may have to adjust your oscilloscope volts/division settings and time/division settings so that the AC input waveform appears on your o-scope display. Describe the appearance of the waveform below.

11 Determine the period of the waveform that appears on your o-scope.

$$p = \frac{\text{\# horizontal divisions}}{1} \times \frac{\text{TIME}}{\text{DIV}} =$$

Now, use the period to determine the frequency of the waveform.

$$f_o = \frac{1}{p} =$$

f_o _____

12 How does the measured frequency of oscillation in Step 11 compare to the calculated frequency of oscillation in Step 4? Explain any significant differences.

13 Power off the 12 V DC source and the o-scope. Set your DMM to measure frequency. Connect the positive lead of the DMM to the collector of Q_1 or to the top end of the tank circuit. Connect the negative end of your DMM to ground. Power on the 12 V DC source and your DMM. Measure and record the frequency.

f_o _____

14 How does the measured frequency of oscillation in Step 13 compare to the calculated frequency of oscillation in Step 4?

15 Have the instructor verify your results.

16 Build the circuit on Multisim®, Electronics Workbench®, or SPICE and then compare the results.

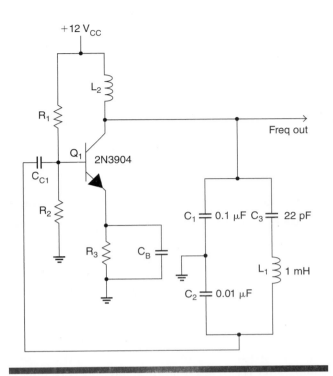

+12 V_{CC}

FIGURE 9-7 Clapp oscillator (© Cengage Learning 2012)

Situation:

What is the frequency of oscillation for the Clapp oscillator in Figure 9-7?

Solution:

$$C_T = \cfrac{1}{\cfrac{1}{C_1} + \cfrac{1}{C_2} + \cfrac{1}{C_3}}$$

$$= \cfrac{1}{\cfrac{1}{0.1\ \mu F} + \cfrac{1}{0.01\ \mu F} + \cfrac{1}{22\ pF}}$$

$$= 21.94\ pF$$

$$f_o = \frac{1}{2\pi\sqrt{LC_T}}$$

$$= \frac{1}{2\pi\sqrt{(1\ mH)(21.94\ pF)}}$$

$$= 1{,}075{,}033.61\ Hz$$

The last LC oscillator we'll discuss is the **Clapp oscillator**. The Clapp oscillator is really just a variation of the Colpitts oscillator. Figure 9-7 shows a Clapp oscillator.

In Figure 9-7, notice that capacitor C_3 has been added to the tank circuit in series with coil L_1. This capacitor reduces the effects of rogue capacitance that occurs between the leads of the other two capacitors, the leads of the coil, and the wires used in the feedback circuit. The Clapp oscillator has a more accurate output frequency than the Hartley or Colpitts oscillators. When calculating the frequency of oscillation for a Clapp oscillator, C_3 will figure into the C_T.

Thus, the Clapp oscillator in Figure 9-7 will generate a frequency of @ 1.07 MHz. The Clapp oscillator's measured frequency should be close to its calculated frequency. Now, we'll look at oscillators whose output frequencies are more precise than LC oscillators.

9-4 CRYSTAL OSCILLATOR

The **crystal oscillator** is a high-precision oscillator that uses a quartz crystal to create a sine wave. Like the Hartley and Colpitts oscillators, crystal oscillators are used in many electronic communications applications including signal generators, radio receivers, and radio transmitters. In addition, the precise output frequency and stability of crystal oscillators makes them ideal for use in applications where timing is critical: computer motherboards, video cards, network cards, and USB flash drives; digital clocks; cell phones and iPods; and measuring and monitoring instruments such as oscilloscopes and pacemakers.

The crystal oscillator's operation is based on **piezoelectricity**. A voltage, or electromotive force, "squeezes" the crystal, distorting its shape, and then the crystal "squeezes" back. (Yes, you can squeeze something out of a rock!) This back and forth contraction and expansion produces vibrations or oscillations (sine waves) of a particular frequency, called the crystal's resonant frequency. A crystal has both a series resonant frequency and a parallel resonant frequency, so it can replace either a series LC circuit or a parallel LC circuit (tank).

The frequency of operation, or resonant frequency, of a quartz crystal (much like the value of a diamond) is determined by several factors including size, shape, cut, and thickness. The thinner the cut of a quartz crystal, the higher its resonant frequency. A typical quartz crystal is cut thinner than paper for use in the MHz range. Most crystals are cut and shaped to a specific frequency.

During the manufacturing process, the crystal is mounted on a stand between two metal plates, all of which are then sealed in a metal box. Figure 9-8 shows a crystal oscillator.

In Figure 9-8(a), the crystal has been removed from its protective box, and you can see how thin it is. Figure 9-8(b) shows a crystal oscillator on the circuit board of an older network interface card. The

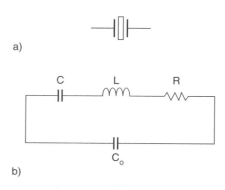

FIGURE 9-9   Crystal oscillator: a) Schematic symbol and b) Equivalent circuit (© Cengage Learning 2012)

frequency stamped on the case—50.000 MHz—indicates the precision of the oscillator, 1/1000 of a MHz. The *NTE* manual lists quartz crystals with frequency tolerances of ±0.003% to ±0.005%! Crystal oscillator metal cases usually have rectangular, square, or oval shapes, but the case shape doesn't indicate the shape of the internal crystal. The internal crystal may be cut into the shape of a circle, a rectangle, or a tuning fork.

Figure 9-9(a) shows the schematic symbol for the crystal oscillator, which resembles the symbol for a capacitor with a box inside it. The schematic designator for crystal oscillators begins with the letter *Y* followed by a number that is usually in subscript form. Sometimes you will see the letter *X* or "XTAL" used.

Figure 9-9(b) shows the equivalent circuit for a crystal oscillator. Notice that it resembles the RLC circuit we studied in AC class that had a resistor, inductor, and capacitor. The only difference is the added capacitor in parallel with the RLC combination, called the static capacitance, or C_o. According to the *NTE* manual, C_o represents the capacitance that exists between the pins of the crystal oscillator, which includes the mounting stand, the metal plates that hold the crystal, and the crystal itself.

Although Alexander M. Nicholson invented the first crystal oscillator at Bell Labs in 1918, his design used a Rochelle salt crystal. George Washington Pierce invented the simplest type of *quartz* crystal oscillator circuit in 1920 using a vacuum tube for amplification. Figure 9-10 shows a solid-state version of the Pierce oscillator.

In Figure 9-10, the crystal determines the frequency of operation. Q_1 is an N-channel JFET that provides amplification. Positive feedback is

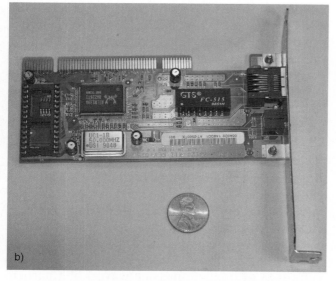

FIGURE 9-8   Crystal oscillator: a) Crystal mounted on stand and b) Network interface card with 50.000 MHz crystal oscillator (Courtesy of Tracy Grace Leleux)

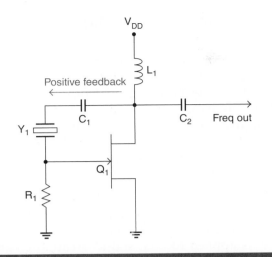

FIGURE 9-10 Pierce oscillator (© Cengage Learning 2012)

overtone, frequencies. The second method uses a quartz crystal in an application with another LC circuit such as the Colpitts oscillator. Both involve additional circuitry.

Quartz crystals can be cut to vibrate at more than one frequency of operation. The natural frequency of a quartz crystal is called its resonant frequency, or fundamental frequency. However, a quartz crystal might also vibrate at a frequency that's a multiple of its fundamental frequency, which is called a harmonic frequency, or overtone. Figure 9-11 shows a circuit that will provide harmonic or overtone frequencies, increasing the frequency output of the crystal oscillator beyond 15 MHz.

In Figure 9-11, crystal Y_1 provides the fundamental frequency of operation, and adjustable coil L_1 allows adjustment of the output frequency. By changing the values of capacitors C_1 through C_4 and coil L_1, the circuit can produce output frequencies ranging from 15 to 65 MHz.

The second method used to increase the frequency of operation of a quartz crystal beyond 15 MHz is to combine a quartz crystal with another LC oscillator circuit. The most common circuit combines a quartz crystal with a Colpitts oscillator. The quartz crystal can either replace the tank circuit of the Colpitts oscillator or be placed in series with the feedback circuit.

Crystal oscillators are the most accurate and stable oscillators on the market. Manufacturers and hobbyists continue to develop new variations and applications for crystal oscillators.

provided through capacitor C_1, and V_{DD} supplies the DC voltage needed to jump start the quartz crystal. The Pierce oscillator works well at the crystal's resonant frequency, and it is used extensively in digital clocks and watches.

We mentioned earlier that the thinner the quartz crystal, the higher its frequency of operation. If a piece of quartz is cut too thin, it could shatter when a voltage is applied to it. For this reason, crystals aren't usually cut to resonate beyond 10 to 15 MHz. Instead, two methods are used to increase the frequency of operation of a quartz crystal oscillator. The first approach involves tuning a quartz crystal to one of its harmonic, or

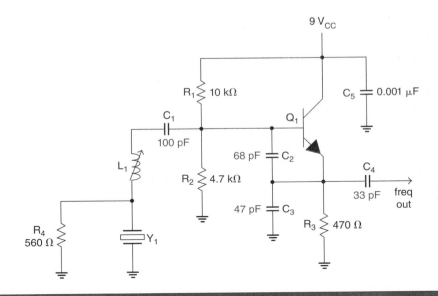

FIGURE 9-11 Crystal oscillator with overtone capability (© Cengage Learning 2012)

CHAPTER SUMMARY

An analog oscillator is an electronic circuit that creates an AC sine wave from a DC voltage source and a tank circuit (or crystal). Four conditions are required for oscillation: a DC voltage source; a tank circuit or crystal; positive, or regenerative, feedback; and amplification.

The tank section is what creates the frequency of oscillation for an oscillator. Oscillator circuits are designed by choosing values of inductance and capacitance that will produce a desired output frequency, which is called the frequency of oscillation, or f_o.

The Hartley oscillator is a sine-wave oscillator that has two coils (or a center-tapped coil) and one capacitor in its tank section. Ralph Vinton Lyon Hartley invented the vacuum tube version of the Hartley oscillator at Western Electric Company in 1915, and the solid-state version of the Hartley oscillator is still used in many applications today, including signal generators and broadcast radio receivers and transmitters. The advantages of the Hartley oscillator include its simplicity of design and its ability to create a fairly constant output voltage over a large range of frequencies. The major disadvantages of the Hartley oscillator involve imprecise frequency accuracy and frequency stability.

The Colpitts oscillator is a sine-wave oscillator that has two capacitors and one coil in its tank section. Edwin H. Colpitts invented the vacuum tube version of the Colpitts oscillator at Western Electric Company in 1919. The solid-state version of the Colpitts oscillator is still used in many applications today, mostly in the electronic communications field. The advantages of the Colpitts oscillator include the simplicity of design and the ability to create a fairly constant output voltage over a large range of frequencies. Like the Hartley, the main disadvantage of the Colpitts oscillator is that the output frequency it generates is not precise, and it is subject to frequency drift. The Clapp oscillator has three capacitors and one coil in its tank section and provides a more accurate output frequency than the Hartley or Colpitts oscillator.

The crystal oscillator is a high-precision oscillator that uses a quartz crystal to create a sine wave. Like the Hartley and Colpitts oscillators, crystal oscillators are used in many electronic communications applications, and their precise output frequency and stability makes them ideal for use in applications where timing is critical such as computers, digital clocks, cell phones, and pacemakers.

The crystal oscillator's operation is based on piezoelectricity, where a voltage, or electromotive force, is applied to a crystal to create oscillations of a particular frequency called the crystal's resonant frequency. The resonant frequency of a quartz crystal depends on the crystal's size, shape, cut, and thickness. The thinner the cut of a quartz crystal, the higher its resonant frequency. A typical crystal is cut thinner than paper for use in the MHz range.

A crystal is mounted on a stand between two metal plates, and the crystal oscillator will behave like an RLC resonant circuit. Alexander M. Nicholson invented the first crystal oscillator at Bell Labs in 1918 using a Rochelle salt crystal, and George Washington Pierce invented the simplest type of *quartz* crystal oscillator circuit in 1920. The solid-state version of the Pierce oscillator is still used today.

Crystals alone can only work up to about 15 MHz without shattering, so two methods are used to increase the frequency of operation of a quartz crystal oscillator. The first approach involves tuning a quartz crystal to one of its harmonic or overtone frequencies. The second method uses a quartz crystal in an application with another LC circuit such as the Colpitts oscillator. Both involve additional circuitry.

Crystal oscillators are the most accurate and stable oscillators on the market, and manufacturers continue to conduct research and development of crystal oscillator technology for future applications.

CHAPTER EQUATIONS

Hartley oscillator frequency of oscillation

(Equation 9-1)

$$f_o = \frac{1}{2\pi\sqrt{L_T C}}$$

f_o = frequency of oscillation in Hz

L_T = total inductance of coils in tank
    (if more than one coil)

C = capacitance of capacitor in tank

Colpitts oscillator frequency of oscillation

(Equation 9-2)

$$f_o = \frac{1}{2\pi\sqrt{L C_T}}$$

f_o = frequency of oscillation in Hz

L = inductance of coil in tank

C_T = total capacitance of capacitors in tank

CHAPTER REVIEW QUESTIONS

Chapter 9-1

1. What is an analog oscillator?

2. List the four requirements for oscillation.

3. Why does an oscillator circuit need amplification?

Chapter 9-2

4. What section of a Hartley oscillator determines the frequency of oscillation?

5. Explain the flywheel effect in a Hartley oscillator.

6. What is the difference between positive and negative feedback?

7. What electronic device discussed in a previous chapter uses negative feedback?

8. Calculate the frequency of oscillation for the tank section of the Hartley oscillator in Figure 9-12.

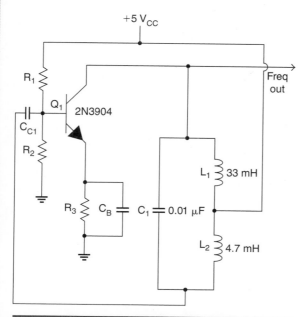

FIGURE 9-12   Hartley oscillator (© Cengage Learning 2012)

9. Calculate the frequency of oscillation for the tank section of a Hartley oscillator if $L_1 = 47$ mH, $L_2 = 100$ mH, and $C_1 = 0.22$ µF.

10. What is one disadvantage of a Hartley oscillator?

Chapter 9-3

11. Calculate the frequency of oscillation for the tank section of the Colpitts oscillator in Figure 9-13.

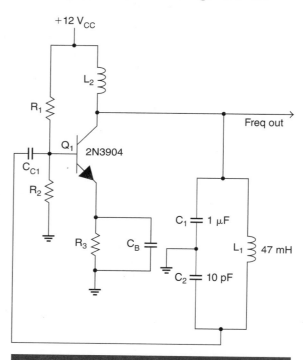

FIGURE 9-13   Colpitts oscillator (© Cengage Learning 2012)

12. Calculate the frequency of oscillation for the tank section of a Colpitts oscillator if $C_1 = 220$ pF, $C_2 = 470$ pF, and $L_1 = 10$ mH.

13. How can one modify a Colpitts oscillator to make it a Clapp oscillator?

14. Which LC oscillator has better frequency stability, a Colpitts oscillator or a Clapp oscillator?

Chapter 9-4

15. Draw the schematic symbol for a crystal oscillator.

16. Explain the concept of piezoelectricity.

17. List and explain the factors that determine the resonant frequency of a crystal.

18. Explain the differences between LC oscillators (Hartley and Colpitts) and crystal oscillators in terms of frequency accuracy and stability.

19. Discuss the two methods used to create additional frequencies from one crystal oscillator.

20. Name a common application for a crystal oscillator.

10

Semiconductor Control Devices

OBJECTIVES   *Upon completion of this chapter, you should be able to:*

- Distinguish between open-loop and closed-loop control systems.

- Draw and label the schematic symbol for a silicon-controlled rectifier (SCR).

- Use the *NTE* manual to determine the forward breakover voltage, gate trigger voltage, and holding current for an SCR.

- Explain the operation of an SCR in DC and AC applications.

- Build and analyze an SCR circuit in a DC application.

- Draw and label the schematic symbol for a diac.

- Draw and label the schematic symbol for a triac.

- Explain the operation of a diac-triac circuit used to control a load.

MATERIALS, EQUIPMENT, AND PARTS

Materials, equipment, and parts needed for the lab experiments in this chapter are listed below:

- *NTE* catalog or Internet access, www.nteinc.com (*Look it up to hook it up.*)
- PC w/Multisim®, Electronics Workbench®, or SPICE.
- DMM with test leads.
- Variable DC voltage source, 0 V–5 V.
- Breadboard and connecting wires.
- C106-series SCR, T106-series SCR, or equivalent SCR.
- Two-terminal LED (any color).
- 270 Ω fixed resistor, 470 Ω fixed resistor, and 1.5 kΩ fixed resistor.
- SPST switch (2).

GLOSSARY OF TERMS

Open-loop control system An electronic control system that has no feedback or self-adjustment capability

Closed-loop control system An electronic control system that has feedback and self-adjustment capability

Sensor An electronic device in a closed-loop control system that monitors some output quantity such as temperature, light, or motion and converts it to an electrical signal for feedback purposes

Controller The "brain" of a closed-loop control system, usually a microprocessor or microcontroller that makes decisions and directs the operation of other devices

Actuator The electronic device in a closed-loop control system that performs some action; examples include the blower motor in a central air-conditioning unit or a heating element in an automatic clothes dryer

Thyristors A family of semiconductor devices used in high-power control and high-speed switching applications

Silicon controlled rectifier (SCR) A three-terminal semiconductor rectifier used for high-speed switching and phase control applications

Repetitive peak reverse voltage (V_{RRM}) The maximum voltage an SCR can sustain in the reverse bias condition

Gate trigger voltage (V_{GT}) The DC voltage value applied between the gate and cathode of an SCR to cause it to conduct

Holding current (I_H) The minimum value of current needed to keep an SCR conducting

False triggering A condition where an unwanted voltage applied to the gate of an SCR causes the SCR to conduct

Diac A two-terminal bi-directional semiconductor rectifier used for high-power, high-speed switching in AC applications

Triac A three-terminal bi-directional SCR used for high-power, high-speed switching in AC applications

10-1 PRINCIPLES OF ELECTRONIC CONTROL

Electronic control is used in many processes and applications, from vending machines to heating and air conditioning to assembly line production to home automation. Many of the semiconductor devices we've studied so far are used in electronic control systems, and this chapter introduces a few more. The focus of this chapter involves semiconductor devices used in high-power control and high-speed switching applications: the silicon-controlled rectifier (SCR), the diac, and the triac.

Electronic control systems are either open-loop or closed-loop. You should recognize these terms from the op amps we built in Chapter 8. An **open-loop control system** has no feedback or self-adjustment capability. A standard automatic clothes dryer is an example of an open-loop system. You put the clothes in the dryer, set it for a certain temperature and cycle, close the lid, turn it on, and walk away. After a certain amount of time, the dryer stops, and you check the clothes. The clothes might be over dried or still damp because the system provides no feedback or self-adjustment. So, an open-loop system means "Check me now and check me later." Other examples of open-loop systems are microwaves, toasters, and circulating fans.

A **closed-loop control system** provides feedback and self-adjustment. Figure 10-1 shows a block diagram of a closed-loop system.

Figure 10-1 is a simplified version of a typical central air conditioning system used in a home or business. The actual air temperature is shown on the far right, and the desired air temperature is shown on the far left. The key devices between these variables are the sensor, comparator, controller, power control and amplification devices, and the actuator. Together, these devices make up the closed-loop system.

In short, an air conditioner removes warm air and cycles it back through a condenser as cool air. The air-handling unit inside the house pulls warm air from the rooms. This warm air travels through ducts to the outside condenser unit, which cools the air. The air-handling unit then reroutes the cool air back inside the house and redistributes it through ducts to the various rooms.

In Figure 10-1, the **sensor** converts the actual air temperature to an electrical voltage, which is then sent to the comparator. Thus, the sensor provides the monitoring and feedback in the closed-loop system. The comparator works like the op amp comparator; it compares the desired temperature (thermostat setting) to the actual temperature and determines the difference. The resulting electrical signal is sent to the controller.

The **controller** is the brain of the system. It is usually a microcontroller board with low-power electronic circuitry that can make "decisions." If the difference between the desired temperature and the actual temperature is greater than zero, then the controller provides the control signals to turn on the **actuator(s)**. The actuators are the motors within the inside air-handling unit and the outside condenser. As long as the actual temperature differs from the desired temperature, the controller will command

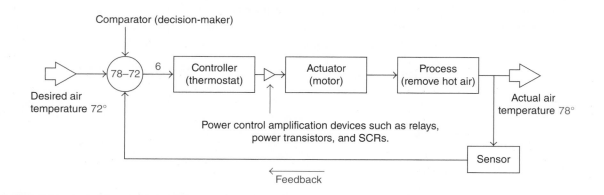

FIGURE 10-1   Block diagram of a closed-loop control system (© Cengage Learning 2012)

the system to keep running. When the desired and actual temperatures are equal, the comparator will send a signal to the controller, which will then shut off the air-handling and condenser units. Thus, the closed-loop control system can be considered "Set it and forget it."

The shortcoming of this central air conditioning system involves the controller. The controller has low-power circuitry, but motors require high amperage. Thus, electronic devices capable of handling high-power and high-speed switching are needed. The small triangle between the controller and the actuator in Figure 10-1 represents these devices, which are the subject of this chapter. **Thyristors** make up the family of semiconductor devices used in high-power control and high-speed switching applications. The most common thyristors are the SCR, diac, and triac.

We've discussed briefly the closed-loop system and the electronic devices that drive the system. For more detailed information on process control, I recommend Chris Kilian's book *Modern Control Technology* from Thomson/Delmar Learning.

10-2 SILICON-CONTROLLED RECTIFIERS (SCRs)

A **silicon-controlled rectifier (SCR)** is a three-terminal rectifier used for high-speed switching and phase control applications. The SCR was invented in 1957 at General Electric by Gordon Hall and Frank W. "Bill" Gutzwiller. Figure 10-2 shows two case styles for the SCR and its schematic symbol.

The SCR on the right of Figure 10-2(a) is the smallest case style of SCR used; it can handle about 30 V and 4 A. The other case style in Figure 10-2(a) can handle up to 400 V and 4 A. Keep in mind that SCRs come in many different case styles and operating specifications. SCRs can resemble power MOSFETs, spark plugs, or even hockey pucks. Again, *Look it up to hook it up*. For example, the *NTE* manual lists the NTE 5563 "hockey puck" SCR as having a 2.9 inch diameter and capable of withstanding a maximum voltage of 1600 V and a maximum current of 1880 A! Yes, over a thousand volts and amps. Now, that's some power rectifier.

The schematic symbol for an SCR is shown in Figure 10-2(b). Notice that the SCR's schematic

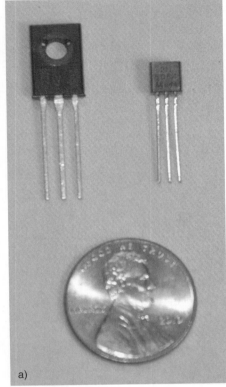

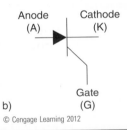

a)

b)

© Cengage Learning 2012

FIGURE 10-2   SCR: a) Two SCR case styles and b) Schematic symbol

symbol looks like the symbol for a diode except for the added gate terminal. This gate terminal plays a critical role in the operation of an SCR.

The internal construction of an SCR resembles that of a diode. Figure 10-3 compares the physical make-up of an SCR to that of a PNP transistor and PN junction diode.

Note that the SCR has two parts P-type material and two parts N-type material; thus, it is like two PN junction diodes connected together. The gate terminal connects to the second P-type material. The SCR will behave like a diode because it passes current only in one direction, from the anode on the far left to the cathode on the far right.

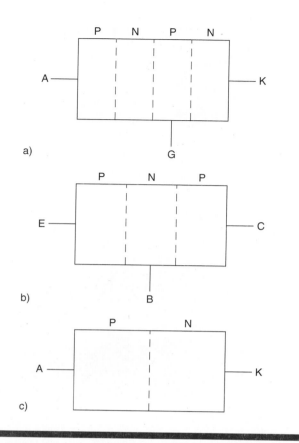

FIGURE 10-3   Construction types: a) SCR, b) PNP transistor, and c) PN junction diode (© Cengage Learning 2012)

Two methods are used to turn on an SCR. If a very large forward bias is applied to the anode and cathode (positive to anode and negative to cathode), then the SCR will start to conduct current. The voltage can be 25 V to 1600 V depending on the SCR. This method is similar to forward biasing a PN junction diode except that a PN junction diode needs only @ 0.2 V to 0.7 V to turn on. This voltage is called the forward breakover voltage $V_{(BO)}$, or repetitive peak forward off-state voltage (V_{DRM}). An SCR can also be rated according to the maximum voltage it can sustain in reverse bias condition: the **repetitive peak reverse voltage (V_{RRM})**. Because this breakover voltage can be dangerously high, the first method is usually not used to turn on an SCR.

The second and most common way to turn on an SCR involves the gate. If a very small DC voltage is applied between the gate and cathode (positive to gate and negative to cathode), then the SCR will start conducting. This tiny voltage can be as low as 0.3 V and as high as 4 V depending on the SCR, and it is called the **gate trigger voltage (V_{GT})**. This is

the preferred method—using a small voltage to turn on an SCR. Once conducting, the SCR itself can be used to turn on a load like a large motor that uses hundreds or even thousands of amps, volts, and watts. *The gate terminal can only be used to turn on an SCR. Once the SCR is conducting current, removing the gate voltage has no effect on the SCR operation.*

Once an SCR is on, it will continue to conduct as long as the current through it remains above a value called the **holding current (I_H)**. Figure 10-4 shows two specification sheets for SCRs NTE 5452 through NTE 5458.

Remember, manufacturers often group similar semiconductors on one spec sheet. The spec sheets show the three major characteristics of SCRs that we've discussed so far. Looking at the first spec sheet in Figure 10-4(a) under the heading "Absolute Maximum Ratings," you can see that the repetitive peak reverse voltage (V_{RRM}) for NTE 5457 is 400 V, and the repetitive peak off-state voltage (V_{DRXM}) for NTE 5457 is also 400 V. On the second spec sheet in Figure 10-4 (b) under the heading "Electrical Characteristics," you can see that the DC holding current has a maximum value of 3 mA.

Figure 10-4(b) also lists the maximum gate trigger voltage (V_{GT}) value of 0.8 V, which is quite small. SCRs that have a small gate trigger voltage are often noted with the label "Sensitive Gate" in the *NTE* manual. These sensitive gate SCRs can be subject to a condition known as **false triggering**; that is, they can be accidentally turned on by an outside voltage source such as an electromagnetic pulse from a nearby transformer. To prevent false triggering, a 1 kΩ resistor is often connected between the gate and cathode. This resistor will drop any unwanted external voltage and prevent the SCR from turning on accidentally. Some SCRs are manufactured with this internal resistor.

Two methods are used to turn off an SCR. In the first approach, current must be removed from the anode (positive) end of the SCR. When this happens, the current through the SCR drops below the holding current (I_H) value, and the SCR turns off.

Another way to turn off an SCR is to apply reverse bias to it; that is, put a negative voltage to the anode and a positive voltage to the cathode. Since alternating current constantly switches directions, you can use AC to both turn on and turn off an SCR.

ELECTRONICS, INC.
44 FARRAND STREET
BLOOMFIELD, NJ 07003
(973) 748–5089
http://www.nteinc.com

NTE5452 thru NTE5458
Silicon Controlled Rectifier (SCR)
4 Amp Sensitive Gate, TO202

Description:
The NTE5452 through NTE5458 are sensitive gate 4 Amp SCR's in a TO202 type package designed to be driven directly with IC and MOS devices. These reverse–blocking triode thyristors may be switched from off–state to conduction by a current pulse applied to the gate terminal. They are designed for control applications in lighting, heating, cooling, and static switching relays.

Absolute Maximum Ratings:
Repetitive Peak Reverse Voltage (T_C = +100°C), V_{RRM}

NTE5452	30V
NTE5453	50V
NTE5454	100V
NTE5455	200V
NTE5456	300V
NTE5457	400V
NTE5458	600V

Repetitive Peak Off–State Voltage (T_C = +100°C), V_{DRXM}

NTE5452	30V
NTE5453	50V
NTE5454	100V
NTE5455	200V
NTE5456	300V
NTE5457	400V
NTE5458	600V

RMS On–State Current, $I_{T(RMS)}$ . . . 4A
Peak Surge (Non–Repetitive) On–State Current (One Cycle at 50 or 60Hz), I_{TSM} . . . 20A
Peak Gate–Trigger Current (3µs Max), I_{GTM} . . . 1A
Peak Gate–Power Dissipation ($I_{GT} \leq I_{GTM}$ for 3µs Max), P_{GM} . . . 20W
Average Gate Power Dissipation, $P_{G(AV)}$ . . . 200mW
Operating Temperature Range, T_{opr} . . . –40° to +100°C
Storage Temperature Range, T_{stg} . . . –40° to +150°C
Typical Thermal Resistance, Junction–to–Case, R_{thJC} . . . +5°C/W

FIGURE 10-4(a) Specification sheets for SCRs, NTE 5452 through NTE 5458 (Courtesy of NTE Electronics, Inc.)

Electrical Characteristics:

Parameter	Symbol	Test Conditions	Min	Typ	Max	Unit
Peak Off–State Current	I_{RRM}	V_{RRM} = Max, V_{DRXM} = Max, T_C = +100°C, R_{G-K} = 1kΩ	–	–	100	μA
	I_{DRXM}		–	–	100	μA
Maximum On–State Voltage	V_{TM}	T_C = +25°C, I_T = 4A (Peak)	–	–	2.2	V
DC Holding Current	I_{HOLD}	T_C = +25°C	–	–	3	mA
DC Gate–Trigger Current	I_{GT}	V_D = 6VDC, R_L = 100Ω, T_C = +25°C	–	50	200	μA
DC Gate–Trigger Voltage	V_{GT}	V_D = 6VDC, R_L = 100Ω, T_C = +25°C	–	–	0.8	V
Total Gate Controlled Turn–On Time	t_{gt}	T_C = +25°C	–	1.2	–	μs
I^2t for Fusing Reference	I^2t	> 1.5msoc	–	–	0.5	A^2sec
Critical rate of Applied Forward Voltage	dv/dt (critical)	R_{G-K} = 1kΩ, T_C = +100°C	–	8	–	V/μs

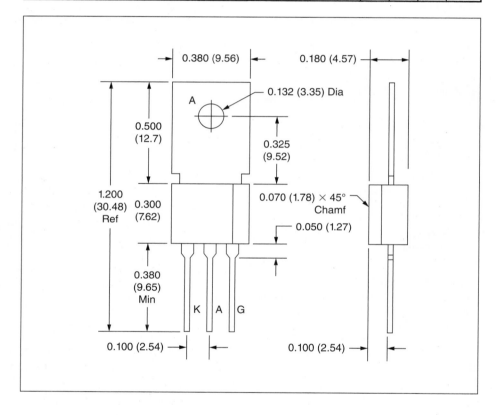

FIGURE 10-4(b) (continued)

10-3 TESTING SCRs

The easiest and most accurate way to test SCRs is to use a semiconductor tester such as a curve tracer. An SCR really needs to be tested under load; that is, with the gate voltage applied and with current flowing through the SCR. However, you can still do a fairly accurate test using your DMM and the diode-setting/six-check method. This method works best for sensitive gate SCRs, which can easily be turned on using the voltage applied by the DMM diode setting.

Remember, an SCR can be considered two PN junction diodes with the gate terminal connected to the P-type material of the second PN junction diode. To test an SCR, set your meter's rotary selector switch to the diode setting. Then touch or clip the red (positive) lead of the meter to the anode terminal of the SCR and touch or clip the black (negative) lead of the meter to the gate terminal of the SCR. The anode and gate are connected to P-type material of two different PN junctions, so the meter should display infinite, which may be represented by the infinity symbol (∞) or "OL" (overload) for a good SCR.

For the second check, touch or clip the red (positive) lead of the meter to the anode terminal of the SCR and touch or clip the black (negative) lead of the meter to the cathode terminal of the SCR. The anode and cathode belong to separate PN junctions, so the meter should display infinite, which may be represented by the infinity symbol (∞) or "OL" (overload) for a good SCR.

For the third check, touch or clip the red (positive) lead of the meter to the gate terminal of the SCR and touch or clip the black (negative) lead of the meter to the anode terminal of the SCR. The anode and gate are connected to P-type material of two different PN junctions, so the meter should display infinite, which may be represented by the infinity symbol (∞) or "OL" (overload) for a good SCR.

For the fourth check, touch or clip the red (positive) lead of the meter to the gate terminal of the SCR and touch or clip the black (negative) lead of the meter to the cathode terminal of the SCR. This forward biases the gate-to-cathode junction of the second PN diode, current flows through the SCR, and the meter should display between 0.6 and 0.7 (volt) for a good SCR.

For the fifth check, touch or clip the red (positive) lead of the meter to the cathode gate terminal of the SCR and touch or clip the black (negative) lead of the meter to the gate terminal of the SCR. This reverse biases the second PN junction, so the meter should display infinite, which may be represented by the infinity symbol (∞) or "OL" (overload) for a good SCR.

For the sixth and final check, touch or clip the red (positive) lead of the meter to the cathode of the SCR and touch or clip the black (negative) lead of the meter to the anode terminal of the SCR. Again, the anode and cathode belong to separate PN junctions, so the meter should display infinite, which may be represented by the infinity symbol (∞) or "OL" (overload) for a good SCR.

In summary, SCR testing using the diode setting and meter leads requires six checks. The results for testing an SCR are summarized below:

1) Positive on anode and negative on gate—display should be infinite or OL.

2) Positive on anode and negative on cathode—display should be infinite or OL.

3) Positive on gate and negative on anode—display should be infinite or OL.

4) Positive on gate and negative on cathode—display should be @ 0.6 to 0.7 (volt).

5) Positive on cathode and negative on gate—display should be infinite or OL.

6) Positive on cathode and negative on anode—display should be infinite or OL.

Note that there is only one numerical reading and five infinite readings for a good SCR. If the ratio is different—three and three or four and two—then the SCR is defective and must be replaced.

The limitation to this diode-setting/six-check method for testing an SCR is that it doesn't necessarily show how the SCR will behave under full-load conditions; that is, when it is powered on and conducting current. Also, this method works only with sensitive gate SCRs. However, it is a fairly reliable test. Curve tracers and specialized SCR testers provide the best accuracy. Again, *when in doubt, swap it out.*

10-4 SCR OPERATION IN A DC CIRCUIT

An SCR can be used in a DC circuit or an AC circuit. Figure 10-5 shows the basic operation of an SCR in a DC circuit. Notice the SCR uses the same schematic designator as a diode: the letter *D* followed by a number. In some circuits, you may see the letters *SCR* followed by a number.

In Figure 10-5, SW1 provides a path for current to the positive end of the load, which could be a light, a motor, or a heavy-duty measuring device used in oilfield drilling. R_3 limits the current through the load. R_1 and R_2 form a voltage-divider circuit, with R_2 providing the gate trigger voltage and with SW2 providing a means to turn on the SCR via the gate terminal.

Figure 10-6 shows the same circuit except that an LED is now the load and the SCR is energized.

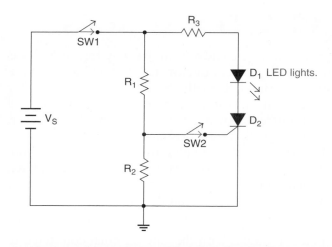

FIGURE 10-6 SCR control of an LED, gate switch on
(© Cengage Learning 2012)

When SW1 and SW2 are closed, the SCR turns on. Current flows through SW1, through R_3, through the LED and SCR, and straight to ground. The LED lights because it has a positive voltage on its anode and a negative voltage on its cathode.

If SW2 is turned off, the SCR stays on, and the LED stays lit. This condition is shown in Figure 10-7. Remember the gate voltage only turns *on* an SCR. However, if SW1 is turned off, the current through the SCR will drop below the holding current value, and the SCR will turn off.

Figures 10-5, 10-6, and 10-7 are simple DC circuits, but they demonstrate the basic operation of an SCR. Let's build an SCR circuit on the breadboard.

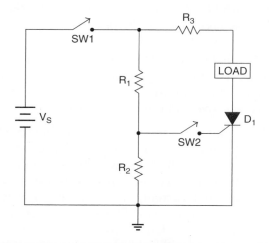

FIGURE 10-5 Basic operation of SCR in a DC circuit
(© Cengage Learning 2012)

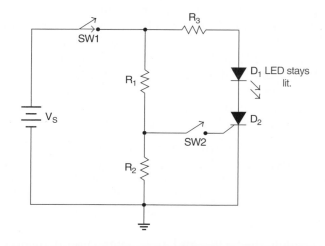

FIGURE 10-7 SCR control of an LED, gate switch off
(© Cengage Learning 2012)

Operation of an SCR in a DC Circuit

Materials, Equipment, and Parts:

- *NTE* catalog or Internet access, www.nteinc. com (*Look it up to hook it up.*)

- PC w/Multisim®, Electronics Workbench®, or SPICE.

- DMM with test leads.

- Variable DC voltage source, 0 V–5 V.

- Breadboard and connecting wires.

- C106-series SCR, T106-series SCR, or an equivalent SCR.

- Two-terminal LED (any color).

- 270 Ω fixed resistor, 470 Ω fixed resistor, and 1.5 kΩ fixed resistor.

- SPST switch (2).

Discussion Summary:

An SCR can be used in a DC circuit to turn on a load such as an LED. A small gate trigger voltage causes the SCR to conduct and will also turn on any load connected in series with the SCR.

Procedure:

SAFETY FIRST. Eye protection should always be worn when working with live voltages. Before powering on a live circuit, always check with your instructor.

1 Write the part number of the SCR at your workstation. Then use the *NTE* catalog or visit the website www.nteinc.com to find and record the NTE replacement number and the diagram number.

Part number _____

NTE number _____

Diagram # _____

2 Draw the SCR pin outline and label the cathode, anode, and gate.

(continues)

LAB ACTIVITY 10-1

(continued)

3 Build the circuit of Figure 10-8 on your breadboard.

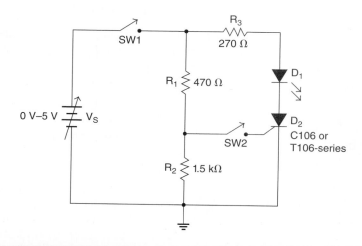

FIGURE 10-8   SCR control of an LED (© Cengage Learning 2012)

4 Connect the positive lead of the variable DC voltage source to one end of SW1.

5 Connect the black lead of the variable DC voltage source, the cathode of the SCR, and a wire from the bottom of resistor R_2 to one point (hole) on the breadboard, which is ground.

6 Ensure the variable DC voltage source is set to its minimum, which should be about 0 volts.

7 Have the instructor check your circuit.

8 Turn on switches SW1 and SW2. Slowly increase the variable DC voltage source until the LED just barely lights.

9 Set your DMM to measure DC volts. Connect the positive lead of your DMM to the gate of the SCR. Connect the negative lead of your DMM to the cathode of the SCR. Power on your DMM and record the measured voltage, which is the gate trigger voltage.

   V_{GT} _____

10 Turn off SW2. Did anything happen to the LED? Why?

11 Turn off SW1. Did anything happen to the LED? Why?

12 Have the instructor verify your results.

13 Build the circuit on Multisim®, Electronics Workbench®, or SPICE and then compare the results.

10-5 SCR CONTROL OF A DC MOTOR

In Lab Activity 10-1, we demonstrated the basic operation of an SCR in a DC circuit. Because of its ability to change AC to pulsating DC, the SCR can also be used in a circuit with an AC voltage source and a DC load. Figure 10-9 shows an SCR used to control a DC motor.

Motor control is one of the most common applications for SCRs. Large horsepower motors used in factories, oil and gas drilling, and machine shops require hundreds of volts to provide the current needed for operation. High-voltage AC can be produced more cheaply and efficiently than high-voltage DC, so AC is usually used to supply the operating voltage for large DC motors.

In Figure 10-9, the voltage source is AC. SW2 provides the gate trigger voltage for the SCR, which is in series with the DC motor. How does this circuit operate?

When the AC voltage source is positive on the top and negative on the bottom and switches SW1 and SW2 are closed as shown in Figure 10-10, the SCR is forward biased, and SW2 provides the gate trigger voltage to turn on the SCR and the motor. Current flows through SW1, through the motor, and through the SCR to the negative side of the AC voltage source.

When the AC voltage source is negative on the top and positive on the bottom as shown in Figure 10-11, the SCR is now in reverse bias—negative to anode and positive to cathode. No current flows through the SCR or the DC motor.

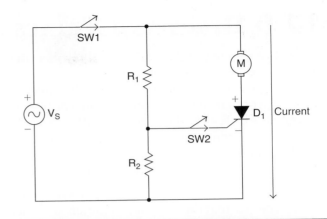

FIGURE 10-10   SCR control of a DC motor during the positive alternation of AC voltage source (© Cengage Learning 2012)

Remember, one way to turn off an SCR is to apply a reverse bias to it. The SCR in Figures 10-10 and 10-11 has provided rectification; it has changed the AC signal to a pulsating DC signal. Since the motor only receives power on the positive alternation of the AC cycle, it is a half-wave rectifier circuit. To make the circuit a full-wave rectifier, another SCR would be needed. This additional SCR would be connected so that it would conduct on the negative alternation.

Figures 10-10 and 10-11 demonstrate the basic principles behind SCR control of a DC motor that uses an AC voltage source. In commercial applications, the circuitry is more complex.

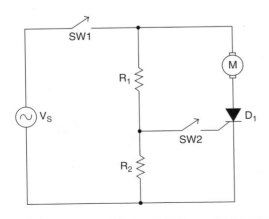

FIGURE 10-9   SCR control of a DC motor (© Cengage Learning 2012)

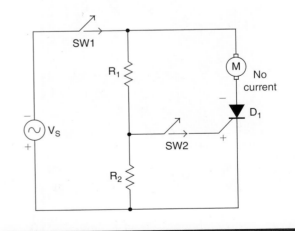

FIGURE 10-11   SCR control of a DC motor during the negative alternation of AC voltage source (© Cengage Learning 2012)

10-6 DIACS AND TRIACS

A **diac** is a two-terminal bi-directional rectifier used for high-power, high-speed switching in AC applications. The diac is similar in operation to a PN junction diode, except that the diac conducts current in both directions. Physically, the diac is smaller than the triac, a three-terminal bi-directional SCR, and it can look like a diode. Figure 10-12(a) shows a diac and two triacs. The diac is the smallest of the three.

Figure 10-12(b) shows the schematic symbol for the diac. Notice that the terminal on the left is labeled MT2, or Main Terminal 2, and the terminal on the right MT1, or Main Terminal 1. This is because there is no designated anode or cathode for the diac. When the breakover voltage is exceeded in either direction, the diac allows current to flow through it. Typically, a diac only needs about 30 V to begin conducting.

As mentioned above, the **triac** is a three-terminal bi-directional SCR used for high-power, high-speed switching in AC applications such as industrial motors and household appliances like dishwashers,

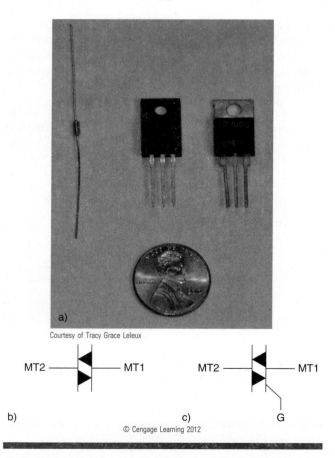

Courtesy of Tracy Grace Leleux

b)   c)
© Cengage Learning 2012

FIGURE 10-12 Diac and triac: a) One diac and two triacs, b) Diac schematic symbol, and c) Triac schematic symbol

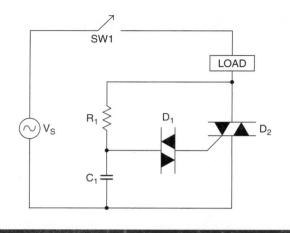

FIGURE 10-13 Triggering a triac with a diac
(© Cengage Learning 2012)

ice makers, and clothes dryers. Like the SCR, the triac needs a gate trigger voltage to turn on. Figure 10-12(c) shows the schematic symbol for the triac. Note that the triac has MT1, MT2, and a gate terminal. When MT1 is positive and MT2 is negative, a small voltage applied to the gate will cause the triac to conduct from MT1 to MT2. When MT2 is positive and MT1 is negative, a small voltage on the gate will cause the triac to conduct from MT2 to MT1.

Diacs and triacs can be used in combination to provide full-wave rectification for industrial motors, light dimmers, and home appliances. Figure 10-13 shows an example of using a diac to trigger a triac to turn on a load such as a washing machine.

When the AC voltage source is positive on the top and negative on the bottom and SW1 is closed as shown in Figure 10-14, capacitor C_1 begins to

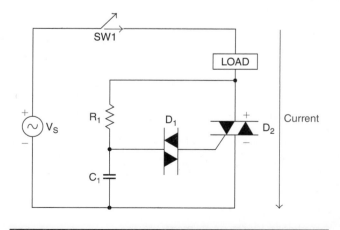

FIGURE 10-14 Triggering a triac with a diac during the positive alternation of the AC voltage source
(© Cengage Learning 2012)

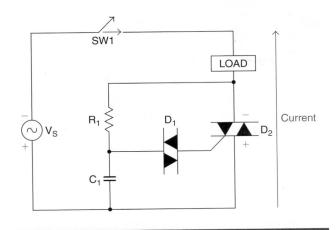

FIGURE 10-15 Triggering a triac with a diac during the negative alternation of the AC voltage source (© Cengage Learning 2012)

charge. When the voltage across C_1 exceeds the breakover voltage of the diac, the diac will turn on. Once the diac starts conducting, it will trigger the gate of the triac. The load will also be energized because it is in series with the triac.

When the AC voltage source is negative on the top and positive on the bottom and SW1 is on as shown in Figure 10-15, capacitor C_1 begins to discharge. The diac will then turn on in the opposite direction, triggering the gate of the triac and keeping the load energized. Thus, for both AC alternations, the load is energized.

Diacs and triacs are used in many more high-power, high-speed applications than we've discussed here. Triacs are used more often than diacs, and they are the semiconductor of choice for AC systems requiring high-power switching.

CHAPTER SUMMARY

Electronic control is used in many processes and applications, from vending machines to heating and air conditioning to assembly line production to home automation. Semiconductor devices used in high-power control and high-speed switching applications include the silicon-controlled rectifier (SCR), the diac, and the triac.

Electronic control systems are either open-loop or closed-loop. An open-loop control system has no feedback or self-adjustment capability. Automatic clothes dryers, microwaves, and toasters are examples of open-loop systems. A closed-loop control system provides feedback and self-adjustment. Examples of closed-loop systems include central air conditioning and automobile cruise control.

An SCR is a three-terminal rectifier used for high-speed switching and phase control applications. SCRs can resemble power MOSFETs, spark plugs, or even hockey pucks. SCRs can handle up to thousands of volts and amps. An SCR can be turned on by applying a very large forward bias to the anode and cathode or by applying a very small gate trigger voltage. SCRs can be turned off by removing the anode current or by applying a reverse bias to the anode and cathode. SCRs are commonly used to trigger large horsepower motors used in factories, oil and gas drilling, and machine shops.

A diac is a two-terminal bi-directional rectifier used for high-power, high-speed switching in AC applications. A breakover voltage of 30 V or less will turn on a diac. The triac is a three-terminal bi-directional SCR used for high-power, high-speed switching in AC applications such as industrial motors and household appliances like dishwashers, ice makers, and clothes dryers. Like the SCR, the triac needs a gate trigger voltage to turn on. Diacs and triacs can be used in combination to provide full-wave rectification for industrial motors, light dimmers, and home appliances, to name a few uses.

CHAPTER REVIEW QUESTIONS

Chapter 10-1

1. Explain the difference between an open-loop control system and a closed-loop control system.

2. Give two examples of an open-loop system and two examples of a closed-loop system.

3. What is considered the brain in a closed-loop control system?

4. What component in a closed-loop system provides monitoring and feedback?

5. Name one example of an actuator.

6. Name the three common electronic devices that belong to the family of thyristors.

7. What general applications typically use thyristors?

Chapter 10-2

8. What is an SCR?

9. Draw and label the schematic symbol for an SCR.

10. What are the V_{DRM}, V_{GT}, and maximum forward current values for a semiconductor with the part number 2N5060?

11. List the two methods used to turn on an SCR.

12. What polarity is the gate trigger voltage for an SCR, positive or negative?

13. True or false. The gate terminal can be used to turn on or turn off an SCR.

14. Explain the importance of holding current I_H to an SCR's operation.

15. What is used to prevent false triggering of an SCR?

16. List the two methods used to turn off an SCR.

Chapter 10-3

17. When testing an SCR using the diode setting, the SCR measures 0 Ω from anode to cathode. Is the SCR good? Why or why not?

Chapter 10-4

18. What letter is used as the schematic designator for SCRs?

19. Name one common application of an SCR.

Chapter 10-5

20. Why is an SCR an effective switching device for controlling large motors?

Chapter 10-6

21. **Draw and label the schematic symbol for a diac.**

22. **Why is a diac often called a bi-directional rectifier?**

23. **Draw and label the schematic symbol for a triac.**

24. **Why is a triac often called a bi-directional SCR?**

25. **See Figure 10-16. Which schematic symbol represents the triac?**

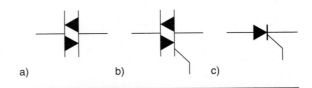

a)   b)   c)

FIGURE 10-16   Thyristor schematic symbols (© Cengage Learning 2012)

Resources

Semiconductor Schematic Symbols

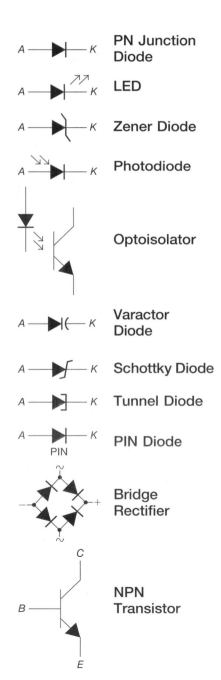

PN Junction Diode

LED

Zener Diode

Photodiode

Optoisolator

Varactor Diode

Schottky Diode

Tunnel Diode

PIN Diode

Bridge Rectifier

NPN Transistor

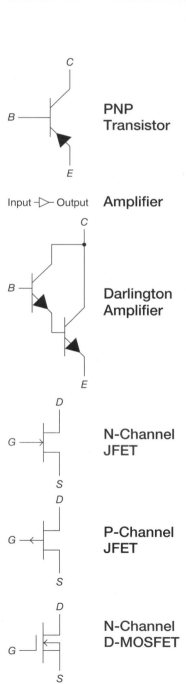

PNP Transistor

Input ▷ Output   Amplifier

Darlington Amplifier

N-Channel JFET

P-Channel JFET

N-Channel D-MOSFET

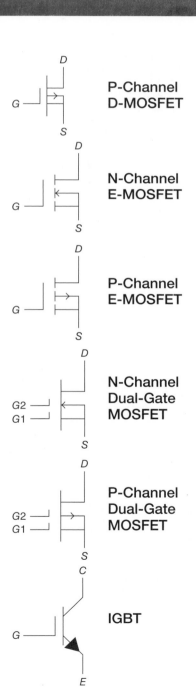

P-Channel D-MOSFET

N-Channel E-MOSFET

P-Channel E-MOSFET

N-Channel Dual-Gate MOSFET

P-Channel Dual-Gate MOSFET

IGBT

Three-Terminal Op Amp

 Five-Terminal Op Amp

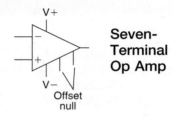 **Seven-Terminal Op Amp**

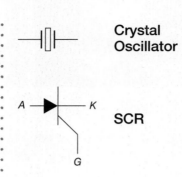 **Crystal Oscillator**

SCR

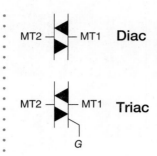 **Diac**

Triac

Glossary

741 op amp The standard eight-pin operational amplifier IC used today in most commercial applications.

A

Actuator The electronic device in a closed-loop control system that performs some action; examples include the blower motor in a central air-conditioning unit or a heating element in an automatic clothes dryer.

Amplifier An electronic device or circuit that takes a small signal (usually AC) and makes it bigger; it can be a power, current, or voltage amplifier.

Amplifier class The category of amplifier that is determined by the way a transistor is biased; that is, the various DC voltages that affect the transistor's operation.

Amplifier configuration The type of transistor scheme determined by the relationship between the AC input signal and the AC output signal.

Analog oscillator An electronic circuit that creates an AC sine wave from a DC voltage source and a tank circuit (or crystal).

B

Barrier voltage (V_B) A force between the P-type material and N-type material of a semiconductor device.

Beta (β) The DC current gain of a bipolar junction transistor that is the result of the collector current divided by the base current; it is also called h_{FE}.

Bias A DC voltage applied to a semiconductor device to turn it on or off.

Bipolar junction transistor (BJT) A three-terminal current-controlled semiconductor device used as a high-speed switch or as an amplifier.

Bypass capacitor A capacitor connected across the emitter resistor in a common-emitter configuration that diverts to ground any AC signal appearing on the emitter terminal; it maintains the voltage gain of a common-emitter amplifier.

C

Clapp oscillator An LC oscillator that has three capacitors and one coil in its tank section; the third capacitor reduces the effects of rogue capacitance, giving the oscillator better frequency stability than the Hartley or Colpitts oscillators.

Class A amplifier An amplifier that is biased at midpoint and has very high fidelity and low efficiency.

Class B amplifier An amplifier that is biased at cutoff and has high fidelity and high efficiency; it usually contains an NPN transistor that reproduces the positive alternation of the AC input signal and a PNP transistor that reproduces the negative alternation of the AC input signal.

Class C amplifier An amplifier that is biased way below cutoff and has low fidelity and excellent efficiency; it usually contains a tank section that restores most of the distorted AC waveform.

Closed-loop control system An electronic control system that has feedback and self-adjustment capability.

Closed-loop voltage gain (A_{CL}) The gain of an op amp with a feedback circuit that contains a feedback resistor R_f and an input resistor R_i.

Colpitts oscillator An LC oscillator that has two capacitors and one coil in its tank section.

Common-anode seven-segment display A seven-segment LED display constructed so the anodes of all the LEDs are linked internally and then connected to two external pins. The cathode of each internal LED is connected to an external pin. To light a particular LED segment, 5 V DC is applied to one of the common-anode pins, and the cathode pin of the particular LED is connected to ground through an external resistor.

Common-base amplifier The BJT amplifier configuration that has the input signal applied between the emitter terminal and the base terminal and the output signal appearing between the collector terminal and the base terminal.

Common-cathode seven-segment display A seven-segment LED display constructed so the cathodes of all the LEDs are linked internally and then connected to two external pins. The anode of each internal LED is

connected to an external pin. To light a particular LED segment, one of the two common-cathode pins is connected to ground through an external resistor, and 5 V DC is applied to the anode pin of the particular LED.

Common-collector amplifier The BJT amplifier configuration that has the input signal applied between the base terminal and the collector terminal and the output signal appearing between the emitter terminal and the collector terminal; it is also called an emitter follower.

Common-drain amplifier The JFET amplifier configuration that has the input signal applied to the gate and the output signal appearing at the source.

Common-emitter amplifier The most popular BJT amplifier configuration that has the input signal applied between the base terminal and emitter terminal and the output signal appearing between the collector terminal and the emitter terminal.

Common-gate amplifier The JFET amplifier configuration that has the input signal applied to the source and the output signal appearing at the drain.

Common-source amplifier The most popular JFET amplifier configuration that has the input signal applied to the gate terminal and the output signal appearing at the drain terminal.

Comparator An op amp configuration that compares applied input voltage(s) and produces a resulting output voltage.

Compensating diodes Diodes used in a class AB amplifier that each provide a voltage drop across the base-to-emitter junction of a transistor to keep it barely on and thus eliminate crossover distortion.

Controller The brain of a closed-loop control system, usually a microprocessor or microcontroller that makes decisions and directs the operation of other devices.

Conventional flow The view that current flows from the positive side of a voltage source to the negative side.

Crossover distortion The undesirable changing of an amplified signal in a class B amplifier that occurs as the task of reproducing the AC input signal is passed on or crosses over from one transistor to the next; it is shown visually by a small horizontal line, or "glitch," between the positive and negative alternations of the AC output signal.

Crystal oscillator A high-precision oscillator that typically uses a quartz crystal to create a sine wave.

Cutoff The operating condition of a bipolar junction transistor in which the transistor has no I_B, no I_C, and is totally off.

D

D-MOSFET A metal oxide semiconductor field effect transistor that can be operated in depletion mode or enhancement mode.

Darlington amplifier or Darlington pair A semiconductor amplifier that uses two common-collector transistors in one housing to provide a large current gain.

DC current gain The current gain of a bipolar junction transistor that is the result of the collector current divided by the base current; it is also called beta (β) or h_{FE}.

Decibel (dB) A unit of measurement that uses the logarithm function and the ratio of two units to calculate an amplifier's gain or loss; it expresses gain in a manageable quantity.

Decoupling capacitor A capacitor connected in series with the DC power supply voltage (V_{CC}) of a common-collector amplifier; it diverts from V_{CC} any unwanted AC on the collector terminal.

Depletion mode A method of operation for a JFET or D-MOSFET in which the transistor is totally on (maximum I_D) when the V_{GS} is 0 V, and by making V_{GS} more negative, the flow of electrons is reduced or depleted.

Depletion region The area that lacks electrons between the P-type material and N-type material of a semiconductor device.

Diac A two-terminal bi-directional semiconductor rectifier used for high-power, high-speed switching in AC applications.

Differential amplifier An op amp configuration that provides an amplified output voltage based on the difference of two input voltages.

Discrete device An individual electronic device such as a resistor, capacitor, or diode.

Distortion The unwanted change in the shape of an AC signal.

Drain-to-source voltage (V_{DS}) Voltage between the drain and source terminals of a FET that causes the flow of drain current.

Dynamic voltage gain The gain of an amplifier with an AC signal applied; the result of the output voltage divided by the input voltage.

E

E-MOSFET A metal oxide semiconductor field effect transistor that can be operated only in enhancement mode.

Efficiency The percentage of DC power that actually makes it to the amplifier's load, such as another amplifier stage or a speaker; it is the result of AC output power divided by the DC input power.

Electron flow The view that current flows from the negative side of a voltage source to the positive side.

Electrostatic discharge (ESD) The release of static electricity that can destroy the gate material in IGFETs; it is often a silent killer, more intense during cool, dry weather.

Emitter bias An amplifier biasing scheme that is not used as often as voltage-divider bias because it requires two power supplies, one positive and one negative.

Emitter follower BJT amplifier configuration where the output voltage tracks or follows the

amplitude and phase of the input voltage; it is also known as a common-collector.

Enhancement mode A method of operation for a D-MOSFET or E-MOSFET where a transistor has a positive V_{GS} that forms a bridge between the drain and source, causing current to flow through the transistor.

F

False triggering A condition where an unwanted voltage applied to the gate of an SCR causes the SCR to conduct.

Feedback resistor (R_f) A resistor connected between the output terminal and the inverting terminal of an op amp that is used to provide negative feedback, thus controlling the gain of an op amp.

Fidelity The ability of an amplifier to "faithfully" or accurately reproduce an AC input signal.

Field effect transistor (FET) A three-terminal voltage-controlled semiconductor device used as a high-speed switch or as an amplifier.

Filtering A process in a power supply where a capacitor or capacitors remove the pulsating or varying portion of a rectified waveform.

Flywheel effect The charging-discharging action between the coil(s) and capacitor(s) of the tank circuit of an LC oscillator that creates a sine wave.

Forward bias condition A situation that occurs in a semiconductor material when the barrier or depletion region breaks down and current flows through the device.

Frequency compensation capacitor A capacitor in an op amp IC used to limit the high-frequency operation of the op amp.

Frequency of oscillation (f_o) The frequency created by the tank circuit of an analog oscillator.

Full-wave bridge rectifier A semiconductor circuit or device made of four diodes that changes AC to pulsating DC by reproducing across a load both alternations of an AC input signal.

Full-wave center-tapped rectifier A semiconductor circuit made of two diodes and a center-tapped transformer that changes AC to pulsating DC by reproducing across a load both alternations of an AC input signal.

Full-wave rectifier A semiconductor circuit that changes AC to pulsating DC by reproducing across a load both alternations of an AC input signal.

G

Gain In an amplifier circuit, the result of the output signal divided by the input signal; that is, how many times bigger the output signal is than the input signal; it can be a power, current, or voltage gain.

Gate-to-source voltage (V_{GS}) Voltage between the gate and source terminals of a FET that controls the flow of drain current.

Gate-to-source cutoff voltage ($V_{GS \ (off)}$) The operating condition of a field effect transistor where the

reverse bias and depletion region are at a maximum, so no drain current or source current flows; it is similar to cutoff in a BJT.

Gate trigger voltage (V_{GT}) The DC voltage value applied between the gate and cathode of an SCR to cause it to conduct.

H

h_{FE} The DC current gain of a bipolar junction transistor that is the result of the collector current divided by the base current; it is also called beta (β).

Half-wave rectifier A semiconductor circuit that changes AC to pulsating DC by reproducing across a load either the positive or negative alternation of an AC input signal.

Hartley oscillator An LC oscillator that has two coils (or a center-tapped coil) and one capacitor in its tank section.

Holding current (I_H) The minimum value of current needed to keep an SCR conducting.

I

IGBT (Insulated Gate Bipolar Transistor) A type of three-terminal insulated gate transistor that is a hybrid of a BJT and MOSFET in terms of construction and operation; it is a fast-switching transistor capable of handling large current and power requirements.

IGFET A category of field effect transistors with extremely high input impedance, characterized by a gate terminal made of an insulating material such as silicon dioxide or hafnium dioxide.

Input resistor (R_i) A resistor connected to an input terminal of an op amp that works with the feedback resistor R_f to control the gain of an op amp.

Integrated circuit An electronic device, often on a semiconductor wafer or chip, that contains several electronic components housed in one package.

Inverting amplifier An op amp configuration that increases and inverts an AC voltage applied to its inverting input terminal; it uses resistors that provide negative feedback to control the op amp voltage gain.

Inverting input terminal One of the input pins of an op amp, represented on a schematic symbol with a negative sign (–). Any AC signal applied to this input terminal will undergo a 180° phase shift by the time it reaches the output terminal; that is, the signal will be inverted or flipped over as well as amplified.

L

Large signal amplifier A transistor circuit that provides a power greater than 0.5 W; it is used in class B and AB amplifiers as the intermediate stage or output stage for audio applications.

Light-emitting diode (LED) A semiconductor device that gives off light when a sufficient voltage is applied to it.

M

Midpoint operation The most desirable operating condition for a transistor amplifier, occurring when the I_C is about one-half the current value at saturation and V_{CE} is about one-half the value of V_{CC}.

MOSFET (Metal Oxide Semiconductor Field Effect Transistor) A type of insulated gate field effect transistor used as a high-speed switch or as an amplifier.

MuGFET (Multiple Gate Field Effect Transistor) A type of field effect transistor constructed of two or more gates that reduce the surface area and thus the overall input capacitance of the transistor.

N

N-channel JFET A junction field effect transistor that controls the movement of electrons through a channel of N-type material.

Negative DC supply voltage terminal A terminal or pin of an op amp that is connected to a negative (−) DC supply voltage source that supplies operating power to the op amp.

Non-inverting amplifier An op amp configuration that increases an AC voltage applied to its non-inverting input terminal; it uses resistors that provide negative feedback to control the op amp voltage gain.

Non-inverting input terminal One of the input pins of an op amp, represented on a schematic symbol with a positive sign (+). Any AC signal applied to this input will be amplified but unchanged in polarity by the time it reaches the output.

NPN A bipolar junction transistor made of two parts of N-type material and one part of P-type material.

O

Offset null terminals Two terminals or pins on an op amp that are used to adjust or calibrate the output of an op amp to 0 volts when it has no input voltage(s) applied; they are used only for applications involving sensitive measurements.

Open-loop control system An electronic control system that has no feedback or self-adjustment capability.

Open-loop voltage gain (A_{OL}) The gain of an op amp without a feedback circuit.

Operational amplifier (op amp) A high-gain amplifier that can increase AC or DC voltages.

Optoisolator A semiconductor device made of an LED-to-photodiode combination that uses light to join an external source to an external load; it is also called an optocoupler.

Organic light-emitting diode (OLED) A light-emitting diode made of carbon atoms encased in metal anodes and cathodes, which in turn are encased in glass plates; this diode is extremely thin, flexible, and has low power consumption.

P

Photodiode A diode that produces a voltage when struck by light; it is used in solar cells and solar panels.

Pin configuration The scheme for identifying the terminals or leads of many electronic devices; it is also called a pin layout or pin-out.

P-channel JFET A junction field effect transistor that controls the movement of holes through a channel of P-type material.

Piezoelectricity The property of certain rock crystals such as quartz and Rochelle salt responsible for creating a sine wave when an electromotive force (voltage) is applied to it.

Pinch-off voltage (V_P) The operating condition of a field effect transistor where increasing the V_{DS} causes no further increase in I_D; it is similar to saturation in a BJT.

PIN diode A specialized diode that behaves like a varactor when reverse biased and like an RF (radio frequency) switch when forward biased, making it useful in high-frequency test probes and in radio communication applications.

PNP A bipolar junction transistor made of two parts of P-type material and one part of N-type material.

Positive DC supply voltage terminal A terminal or pin of an op amp that is connected to a positive (+) DC supply voltage source that supplies operating power to the op amp.

Positive feedback A form of feedback in which the feedback signal is in phase with the input signal, which increases the gain of an amplifier circuit.

Power supply The unit in most electronic systems that provides the voltages to drive the subsystems.

Push-pull An amplifier circuit that uses two transistors working together; one conducts only during the positive alternation of an AC input signal and the other conducts only during the negative alternation of an AC input signal.

R

Rectifier A semiconductor device or circuit that changes AC to pulsating DC.

Regulation A process in a power supply in which an electronic device provides a steady DC voltage to the load.

Repetitive peak reverse voltage (V_{RRM}) The maximum voltage an SCR can sustain in the reverse bias condition.

Resonant frequency (f_r) The frequency of a tuned circuit in an amplifier that is used at the input of a radio receiver or at the output of a radio transmitter.

Reverse bias condition A situation that occurs in a semiconductor material when the depletion region increases and no current flows through the device.

S

Saturation The operating condition of a bipolar junction transistor where the transistor is totally on, the I_C is at maximum, and increasing the I_B won't increase the I_C anymore.

Schottky diode A semiconductor device that has a low turn-on/turn-off voltage, making it useful as a very high-speed electronic switch; it is also known as a hot-carrier diode.

Sensor An electronic device in a closed-loop control system that monitors some output quantity such as temperature, light, or motion and converts it to an electrical signal for feedback purposes.

Seven-segment LED display A semiconductor device that contains seven or more LEDs constructed in a sealed unit; it is used to display numbers and letters in applications such as calculators and automobile instrument panels.

Silicon-controlled rectifier (SCR) A three-terminal semiconductor rectifier used for high-speed switching and phase control applications.

Small signal amplifier A transistor circuit that amplifies voltage signals in the µV to mV range and up to about 0.5 W of power; these small voltages are found at the input of radio receivers, cassette tape heads, vinyl record players, and the preamplifier stage for guitar amplifiers.

Solid-state technology The science involving the movement of electrons (electricity) through a solid piece of semiconductor material, typically silicon or germanium.

Static voltage gain The calculated gain of an amplifier when no AC signal is applied.

Summing amplifier An op amp configuration that provides an amplified output voltage based on the sum of two or more input voltages.

Surface mount technology (SMT) The manufacturing process of printed circuit boards where the pins or leads of an electronic device are soldered directly on the surface or to one side of the board.

T

Through-hole technology The manufacturing process of printed circuit boards where the pins or leads of an electronic device are inserted directly into pre-drilled holes in a circuit board and then soldered on the opposite side to circular pads.

Threshold voltage ($V_{GS\ (Th)}$) The voltage level at which an E-MOSFET forms a bridge across its channel and starts conducting current.

Thyristor A family of semiconductor devices used in high-power control and high-speed switching applications.

Transconductance The amplification rating for a field effect transistor that is the result of a change in I_D divided by a change in V_{GS}; it shows how a change in V_{GS} causes a change in I_D.

Transformer An electrical device typically used in a power supply to reduce or step-down the incoming line voltage.

Transistor A three-terminal semiconductor device used as a high-speed switch or as an amplifier.

Triac A three-terminal bi-directional SCR used for high-power, high-speed switching in AC applications.

Tunnel diode A semiconductor device that provides high-speed switching, making it useful in communication systems, particularly high-frequency applications such as local oscillators for UHF TV tuners; it is also known as the Esaki diode.

V

V_{CC} The main DC supply voltage for a BJT amplifier; it is a positive voltage applied across the entire transistor with respect to ground.

V_{DD} The main DC supply voltage for a FET amplifier; it is a positive voltage applied across the entire transistor with respect to ground.

Vacuum tube technology The science concerning a vacuum tube, an electronic device in which electrons (electricity) move through a low-pressure space, or vacuum.

Varactor A semiconductor device that varies its capacitance according to an applied varying reverse bias voltage and is used in automatic tuning applications; it is also called a varicap.

Voltage-divider bias An amplifier biasing scheme where the main DC power supply voltage is divided among various resistors to turn on a transistor so that it's ready for amplification.

Voltage doubler An electronic circuit that uses two diodes and three capacitors to double the peak output voltage of an AC voltage source while changing it to pulsating DC.

Voltage follower An op amp configuration that has a wire connected from the output terminal to the inverting terminal input, thus limiting the op amp's A_V to 1. The voltage follower, like the BJT emitter follower and the FET source follower, is used only to match a high impedance input stage to a low impedance output stage.

Voltage multiplier An electronic circuit that uses a combination of diodes and capacitors to change AC to pulsating DC *and* increase the output voltage of an AC voltage source.

Voltage regulation The change in output voltage of a power supply from a no-load to full-load condition.

Voltage tripler An electronic circuit that uses three diodes and three capacitors to triple the peak output voltage of an AC voltage source while changing it to pulsating DC.

Z

Zener diode A semiconductor device that conducts when a reverse bias voltage applied to its terminals exceeds a certain value called the V_Z; once conducting, a Zener maintains a steady voltage across its terminals, making it useful as a voltage regulator in power supplies.

Index

spec sheets
    overview, 13
    samples
        diode, 14
        E-MOSFET, 206
        N-channel JFET, 177–178
        op amp, 223–224
        SCR, 290, 291–292
        transistor, 103
static electricity, 200–201
static voltage gain
    of common-emitter amplifiers,
        126–128
    of common-source amplifiers,
        191–194
    defined, 121
    of differential amplifiers,
        245–246
    of inverting amplifiers, 237
    of non-inverting amplifiers, 229
summing amplifiers
    discussion, 246–249
    lab activity, 251–253
surface mount technology, 81
Swartzel, Karl D., Jr., 221
switches
    BJTs as, 185
    diacs and triacs, 298–299
    diodes as, 48
    E-MOSFETs, 205
    JFETs as, 184–185, 187–188
    MOSFETs as, 200
    RF, 48
    SCRs. See SCRs
    transistors as, 113–116

T

tank sections, 163–164, 263.
    See also oscillation and
    oscillators

threshold voltage, 205
through-hole technology, 81
thyristors, 289, 298–299.
    See also SCRs
total gain in multistage
    amplifiers, 146
transconductance, 191–194
transformers, 54, 81–82.
    See also rectification and
    rectifiers
transistors. See also BJTs;
    FETs; oscillation and
    oscillators
    as amplifiers. See amplifiers
    basics, 101–102
    biasing, 111–113
    as high-speed switches, 113–114
    identifying, 102–104
    lab activity, 115–116
    testing, 104–106
triacs, 298–299
troubleshooting tips
    for amplifiers, 129
    for DMMs when testing diodes, 9
    for ICs, 44
    for MOSFETs and static electricity,
        201
    for operational amplifiers, 255
    for power supplies, 83
    for seven-segment displays, 34
tunnel diodes, 48

V

vacuum tube technology, 5–6
varactors, 47
V_{CC}, 123–124
V_{DD}, 189–190

voltage. See also AC; average
        voltage; DC; peak voltage
    barrier, 6–7
    difference, 245–246
    gate trigger, 290
    gate-to-source cutoff, 184
    pinch-off, 184
    repetitive peak reverse, 290
    RMS, 227
    threshold, 175, 205
voltage doublers
    discussion, 87–88
    lab activity, 89–90
voltage followers, 137, 220,
    228
voltage gain, 123, 144–145
voltage multipliers, 87–88, 91
voltage reduction stage, 54
voltage regulation
    lab activity, 41–42
    for power supplies, 82–83
    using Zener diodes, 37–40
voltage triplers, 91
voltage-divider bias
    in amplifiers, 123–124
    in JFETs, 189–190

W

Web sites. See Internet alerts
Widlar, Bob, 221

Z

Zener diodes
    lab activity, 41–42
    main discussion, 37–40
    in MOSFETs, 200
    in power supply example, 81–82